Springer Proceedings in Business and Economics

More information about this series at http://www.springer.com/series/11960

Victoria L. Lemieux
Editor

Building Trust in Information

Perspectives on the Frontiers of Provenance

Springer

Editor
Victoria L. Lemieux
School of Library, Archival and Information
Studies (Information School)
University of British Columbia
Vancouver
British Columbia
Canada

ISSN 2198-7246 ISSN 2198-7254 (electronic)
Springer Proceedings in Business and Economics
ISBN 978-3-319-40225-3 ISBN 978-3-319-40226-0 (eBook)
DOI 10.1007/978-3-319-40226-0

Library of Congress Control Number: 2016947454

Printed on acid-free paper

This Springer imprint is published by Springer Nature
The registered company is Springer International Publishing AG Switzerland

Foreword

To the uninitiated, "provenance" may sound like something that is only a dull and technical topic, but this is a misapprehension. The issues of provenance are pervasive, important, substantive, and interesting. Every television detective drama has a provenance story at its core, namely, the search for criminal evidence and the piecing together of the evidence in a story that explains how the crime occurred. On television, the provenance questions are always resolved and the bad guys are always caught; in the real world, it's not so simple.

I come to this project not with any particular expertise on provenance—the authors of the chapters that follow offer those skills in abundance—but as an end user of the enormous quantities of data emerging from the elaborate ecosystem of financial firms and their regulators and customers. Among many other things, data represent the primary raw material for my own area of monitoring and analyzing financial stability and systemic risk. It is from that perspective that I approach the issues of data provenance considered in this volume.

Finance involves questions about the commitment of economic resources over time. Who will make decisions for the Acme Corporation between now and the next annual meeting? Who should receive repayment of principal when the bond matures? May the mortgagor sublet the house during the life of the loan? In some cases, these commitments take the form of precise contractual promises of future behavior. In other cases, typically where the decision space is too large or the uncertainties too great, decisions are not pre-specified but rather are delegated to trusted agents. Regardless, the intertemporal nature of these questions implies that those making the ex ante commitments may not be around later—or their recollections or loyalties or intentions may not be trusted at that time. Instead, we write things down. We then rely on an elaborate edifice of operational systems to recall those records, together with analysis of laws and institutional frameworks to interpret and enforce them. If we are to trust this edifice, founded on data and analysis, we must understand how it came to be—where the data originated, how they were interpreted, and how conclusions arise from the combination of the data with analysis.

The recent financial crisis illustrates the importance of provenance in financial systemic risk analysis. After more than a year of foreshocks, large and small, the final implosion of the financial sector during the recent crisis began with the failure of Lehman Brothers in September 2008. It is not difficult to point to anecdotes of the deleterious consequences of poor management of data provenance in finance. One might debate which data provenance challenge was the most important, but the Lehman failure was surely the most dramatic. Lehman's London-based subsidiary, Lehman Brothers International Europe (LBIE), commingled its customers' collateral, in a legal regime that tolerated this practice. Commingling meant that collateral was held and tracked in a common account shared by multiple beneficial asset holders. Lehman's failure triggered a mad dash by customers to recover their assets. With billions of dollars' worth of collateral, much of which had been pledged and re-pledged several times, it is an enormous informational and computational task to resolve the questions of identifying the priority claimants and how much they are owed, especially given that the insolvency implied that losses almost surely would be imposed. The mess was turned over to the bankruptcy court in London, which is charged with the task of unraveling the provenance of the contending claims in this portfolio. The court is still sorting out the details almost eight years later. In the meantime, the (formerly) liquid assets of Lehman's counterparties, including many of the key nodes in the financial system, were frozen, setting off a chain reaction of defaults.

In sum, provenance may be a technical subject. Dull it is not. Thus, this book provides foundational reading for anyone looking to expand his or her understanding of provenance.

Washington, DC, USA
May 2016

Mark D. Flood

Preface

The genesis of this volume was a workshop on distributed data analytics held in October of 2014. At that workshop, discussion turned to the fact that organizations are increasingly dependent on information stored and processed by distributed, heterogeneous systems combining both machine and human intelligence for critical decision-making and action. These environments are dynamic in nature and are becoming ever more complex. In such an evolving and complex information landscape, knowing how information is derived—i.e., its provenance—is of great importance in determining the trustworthiness of that information and any decisions and actions taken on the basis of it.

As an archival scientist, provenance has always been at the heart of my interests. However, provenance is also the concern of many other fields as well—law, library and information science, computer science, and visual analytics, to name those addressed in this volume. At the workshop on distributed data analytics, it was clear that diverse fields—all interested in the challenges of conceptualizing, capturing, representing, and using provenance—in many cases worked without knowledge of the theories, research, and practice of the other fields. Thus, the idea was born of hosting a workshop to bring together a group of researchers and practitioners from several different domains to forge an interdisciplinary and multidisciplinary understanding of provenance. This volume is the output of that workshop, held in May of 2015, and subsequent reflections that emerged from it.

The volume begins with a synthesis of the ideas contributed by workshop participants in pre-workshop statements about their interest in and work on provenance, as well as from their interactions at the workshop. The following sections group together chapters presenting thinking and research on provenance of several workshop participants covering diverse domains of interest: archival science, library and information science, computer science, and cognitive science through the lens of visual analytics. Although these chapters do not represent all possible perspectives on provenance, the hope is that the perspectives found in this volume will contribute to an enriched interdisciplinary and multidisciplinary understanding of provenance that can be used to inform new research needed to address the enormous challenge of capturing, representing, and analyzing provenance information in the context of increasingly distributed, heterogeneous information ecosystems.

Vancouver, BC, Canada
August 2016

Victoria L. Lemieux

Acknowledgments

V.L.L. would to thank the Social Sciences and Humanities Research Council of Canada, the US National Institute of Science and Technology, the Research Data Alliance, and InterPARES Trust for providing funding and logistical support for the workshop that led to the development of this volume. Without their generous support, the workshop would not have taken place and this publication would not have been possible. My thanks go to each of the participants of the workshop for sharing ideas and knowledge on the subject of provenance. Finally, I would like to express my appreciation to the staff at Springer for the excellent editorial assistance.

Contents

About the Contributors

Simon Attfield is a Senior Lecturer at the Interaction Design Centre, Middlesex University, and coordinator of the UK Visual Analytics Consortium. Simon's research lies in the area of understanding how people work with information, processes involved in individual and collaborative sensemaking, and implications for interactive systems design. He is co-author of the book *Interacting with Information*, part of the Morgan & Claypool series of Synthesis Lectures on Human-Centered Informatics. He has conducted field-based user-studies at national news organizations, legal firms, and healthcare settings. He has also consulted to news, legal, and medical information providers. He received a B.A. in Philosophy and a B.Sc. in Experimental Psychology from Sussex University and a Ph.D. in Human Computer Interaction from University College London.

Lucie Burgess is an Associate Director for Digital Libraries at the Bodleian Libraries, University of Oxford, and a Senior Research Fellow at Hertford College, Oxford. Lucie leads the Bodleian Digital Library team of 40 staff to deliver high-quality information services and digital research services for the academics and students of the University of Oxford and is a member of the Bodleian Executive. Current projects include *The Alan Turing Institute Symposium on Reproducibility for Data-Intensive Research,* funded by the Alan Turing Institute; *Ensuring a Digital Future for the Past: Shared Digital Preservation for Cambridge and Oxford,* funded by the Polonsky Foundation; *and Giving Researchers Credit for Their Data,* funded by Jisc. Lucie is a member of Oxford University's IT Committee and Digital Strategy Committee, is a board member of the Digital Preservation Coalition, and is the Jisc representative to the ArXiv.org member advisory board and scientific advisory board. From 2007 to 2014 Lucie worked at the British Library, where as Head of Strategy and Planning she led the development of the British Library's 2020 Vision. She also led the UK Legal Deposit libraries' efforts to extend legal deposit to the digital domain. Lucie has also worked in publishing and business development and strategy at United Business Media, an FTSE-250 information services company, and for Arthur Andersen Business Consulting. Lucie began her career working with the United Nations Framework Convention on Climate Change secretariat in Bonn,

Germany. Lucie has a master's degree in Physics from Hertford College, University of Oxford.

Adrian Cunningham is the Director of Digital Archiving at Queensland State Archives. He has worked at the National Archives of Australia, the Queensland State Archives, and with the International Council on Archives (ICA) to promote collaboration and innovation on strategies, frameworks, standards, tools, and guidelines for electronic records and information management. His leadership and initiative led to the development of the ICA's Principles and Functional Requirements for Records in Electronic Environments—ICA-Req. In addition, Adrian Cunningham has participated on international standards development committees and the development of PARBICA's Recordkeeping for Good Governance Toolkit. In addition to his full-time work and participation on various international committees, he has written numerous articles that have been translated and published internationally. He has also shared his knowledge and experience to the international community through seminars presented in over 17 countries for both English and non-English-speaking audiences.

Robert Fields is an Associate Professor in Computer Science in the Interaction Design Centre at Middlesex University. His work explores peoples' interactions with technology in domains including aviation, command and control, novel displays, design education, and data visualization for mental health. Dr. Fields' work includes empirical studies of complex work environments, the design of novel interactive technologies and devices, and methods to support analyses of usability, human error, and safety. He teaches Human Computer Interaction and Interaction Design.

Victoria L. Lemieux is an Associate Professor in Archival Science and a Faculty Associate at the Peter Wall Institute for Advanced Studies and the Institute for Computers, Information and Cognitive Systems at the University of British Columbia in Vancouver, Canada. Prior to her academic career, she previously held positions as a professional archivist, records manager, and risk manager within the public and private sectors. She has also consulted for the United Nations, the Commonwealth Secretariat, and the World Bank. While at the University of British Columbia, in addition to her teaching, she served for 2 years as Director of the Media and Graphics Interdisciplinary Centre (2012–2014) during which time she was instrumental in setting up UBC's Digital Salon as a locus for the Digital Humanities research. She holds a B.A. (Honours) from the University of Toronto, a master's in Archival Studies from the University of British Columbia, and a Ph.D. from University College, London (Archival Studies, 2002), which focused on the information-related causes of the Jamaican Banking Crisis. Since 2005, she has also been a Certified Information Systems Security Professional (CISSP). She has won several awards and distinctions for her research and professional contributions, including the 2015 Emmett Leahy Award, which recognizes outstanding contributions to the records profession and a 2015 World Bank Big Data Innovation Award for her work on the application of big data analytics to the analysis of social media in Brazil. Her current research is focused on risk to the availability of trustworthy records, in particular in financial contexts, and how these risks impact upon transparency, financial stability, public accountability, and human rights.

Bertram Ludäscher is a professor at the Graduate School of Library and Information Science (GSLIS) at the University of Illinois, Urbana-Champaign, and the Director of the Center for Informatics Research in Science and Scholarship (CIRSS). He also holds faculty affiliate appointments at the National Center for Supercomputing Applications (NCSA) and the Department of Computer Science (CS). Prior to joining UIUC, he was a professor at the Department of Computer Science and at the Genome Center at the University of California, Davis. His research interests span several areas in the data-to-knowledge life cycle, from modeling and design of databases and workflows to knowledge representation and reasoning. His current research focus includes theoretical foundations of provenance and practical applications, e.g., to support automated data quality control and workflow-supported data curation. He is a member of the DataONE leadership team, focusing on data and workflow provenance. Until 2004 Ludäscher was a research scientist at the San Diego Supercomputer Center (SDSC) and an adjunct faculty at the Computer Science and Engineering Department at UC San Diego. He received his M.S. (Dipl.-Inform.) in computer science from the Technical University of Karlsruhe (K.I.T.) and his Ph.D. (Dr.rer.nat.) from the University of Freiburg, both in Germany.

Giovanni Michetti is an Assistant Professor in Archival Science at Sapienza University of Rome. He has also taught for some years at the University of British Columbia and the University of Urbino. His research area is focused on contemporary and digital archives. His main research interests are records management, archival description, and digital preservation. He has been involved in international initiatives on digital preservation, including ERPANET (Electronic Resource Preservation and Access Network) and CASPAR (Cultural, Artistic and Scientific Knowledge for Preservation, Access and Retrieval), both funded by UE, and InterPARES (International Research on Permanent Authentic Records in Electronic Systems), funded by the Social Sciences and Humanities Research Council of Canada (SSHRC). He is heavily involved in standardization processes as the Chair of the Subcommittee "Archives and Records Management" and Vice-Chair of the Technical Committee "Documentation and Information" in UNI (the Italian Standards Organization). He is also the Italian representative in a number of ISO Working Groups on records management. He has authored the Italian version of EAD (Encoded Archival Description) and OAIS (Open Archival Information System).

Paolo Missier is a Lecturer in the School of Computing Science at Newcastle University, which he joined in 2011. He has over 20 years of experience in Computer Science development, research, and research management. His core expertise is in

Information Management, with applications mainly to e-science, and more recently in Data Science. He is also one of the contributors to the W3C PROV-O Ontology.

Joe Tennis is an Associate Professor and Director of Faculty Affairs at the University of Washington's School of Information, as well as an Adjunct Associate Professor of Linguistics and a member of the Textual Studies faculty at UW. He is also the President of the International Society for Knowledge Organization (2014–2018), and an Associate Member of the Peter Wall Institute for Advanced Studies at the University of British Columbia. He has been an occasional visiting scholar at the State University of São Paulo since 2009. He is the Managing Editor for Advances in Classification Research Online and on the editorial board for *Library Quarterly* (USA), Knowledge Organization (Germany), *Scire: Representación y Organización del Conocimiento* (Spain), Informatio (Uruguay), and Zagadnienia Informacji Naukowej (Poland). He is also a member of the Dublin Core Usage Board and Oversight Committee (an international standards body that works toward the implementation and maintenance of interoperable metadata). He has been active in the InterPARES (International Research on Permanent Authentic Records in Electronic Systems) research project (working on digital records preservation) since 2005 and currently serves as an advisor and researcher on metadata issues. His research has been funded by Microsoft, Institute of Museum and Library Services (IMLS), and the Social Sciences and Humanities Research Council of Canada (SSHRC). He holds a B.A. in Religious Studies from Lawrence University. He received his M.L.S. from Indiana University and an Sp.L.I.S. in Book History, and the Ph.D. in Information Science from the University of Washington. He works in classification theory, the ethics and aesthetics of information organization labor, the versioning of classification schemes and thesauri, subject ontogeny, information provenance, authenticity metadata, and the comparative discursive analysis of metadata creation and evaluation, including archival metadata, both contemporary and historical. In 2013 he won the ALISE/Bohdan S. Wynar Research Paper Award, for "The Strange Case of Eugenics: A Subject's Ontogeny in a Long-Lived Classification Scheme and the Question of Collocative Integrity," in *Journal of the American Society for Information Science and Technology* 63, 7 (2012): 1350–1359.

Ken Thibodeau is a Guest Scientist, Information Technology Laboratory, National Institute of Standards and Technology (NIST), where he has contributed to the development of ISO/IEC 23000–15 standard, Multimedia Preservation Application Format. Previously, he had a major role in the development of DoD 5015.2, the Electronic Records Management Application (RMA) Design Criteria Standard, and conceptualization of ISO 14721, the Open Archival Information System. In conjunction with computer scientists and engineers at the San Diego Supercomputer Center, he established a conceptual architecture for the persistent preservation of electronic records. This architecture included an archival system that is independent of its information technology infrastructure so that hardware or software components can be replaced with minimal impact on the system as a whole and negligible impact on the records preserved in it. All of this work culminated in the design, development, and successful roll out of the Electronic Records Archives (ERA) Program of the National Archives and Records Administration (NARA). The success of this development has enabled NARA to take on the challenge of ensuring the preservation and providing access to more than 100 terabytes of electronic presidential records of George W. Bush, as well as an additional 300 terabytes of historically valuable electronic records from Congress, federal courts, and agencies.

Caroline Varga is an Associate at Seetru Ltd., Bristol, UK, and L.E.K. Consulting. Previously she has held internships at ExxonMobil, CIC Investor Services Ltd., Deloitte China, Morgan Stanley, and the UK Health Protection Agency. She holds a Master's of Engineering (M.Eng.) and a First in Chemical Engineering from the University of Cambridge and is a Rosemary Murray Scholar.

Margaret Varga is the CEO of Seetru Ltd., Bristol, UK, a company focused on development of visual analytics technologies, and a visiting fellow at Oxford University, Oxford, UK. She is formerly a QinetiQ Fellow. She joined QinetiQ's precursor organization, the Royal Signals and Radar Establishment, after she graduated from Cambridge University. Her current research interests are in visualization, visual analytics, uncertainty analysis, Bayesian reasoning, pattern processing, image processing, content-based image compression, and decision support. She also has an interest in medical image processing, and she developed the world's first automated breast cancer grading system. She is the UK National Leader on the NATO IST-85/RTG41 "Interactive Visualisation of Network Dynamics" and has been UK National Leader on previous NATO IST Visualisation groups since 1996. She was Chairman of the group for 5 years. Dr. Varga has over a hundred publications and is co-inventor of seven patents.

Ashley Wheat is a Ph.D. candidate working in the Interaction Design Centre at Middlesex University. His research seeks to develop a theoretical framework for the analysis of sensemaking through the lens of distributed cognition. During his research, he has carried out studies of military intelligence analysis and worked on projects in the areas of sensemaking, visual analytics, and provenance. Ashley completed a B.Sc. in Multimedia Computing at Middlesex University in 2012. He won a university scholarship to begin his Ph.D. research in late 2013.

Part I
Synthesis

Provenance: Past, Present and Future in Interdisciplinary and Multidisciplinary Perspective

Victoria L. Lemieux and the imProvenance Group

Abstract This chapter presents a multi- and interdisciplinary synthesis of ideas about the definition and theoretical conceptualization of provenance, drawing from disciplines such as archival science, law, computer science, library and information science, and visual analytics. Through the lens of these distinct domains, the chapter explores different purposes served by provenance; various ways that diverse fields are capturing, representing and using provenance information; provenance standards and specifications, and a range of open research challenges relating to theorizing about provenance and capturing, representing and using provenance information in increasingly distributed, heterogeneous information eco-systems combining machine and human intelligence. From this blending of perspectives on provenance from different disciplines and 'interdisciplines', a rich picture emerges of provenance as a dynamic construct and evolving focus of research.

Keywords Metadata • Provenance • Sense-making • Trust • Trusted computing

1 Introduction

The concept of provenance has for many decades been a focus of archival discourse and is, indeed, the basis of a core archival principle (*Respect des Fonds*, or the Principle of Provenance), having been first prescribed for use in Denmark in instructions for the commission on the arrangement of financial archives [1]. In

---

Membership in the imProvenance Group is fluid, but the core group of individuals who contributed to the development of this synthesis comprise: Lucie Burgess, Adrian Cunningham, Ken Cavelier, David Dubin, Luciana Duranti, Paolo Missier, Bertram Ludäscher, Corinne Rogers, Joe Tennis, Ken Thibodeau, Margaret Varga and Ashley Wheat.

V.L. Lemieux (✉)
The University of British Columbia, Vancouver, BC, Canada
e-mail: vlemieux@mail.ubc.ca

the imProvenance Group
The University of British Columbia, Canada

V.L. Lemieux (ed.), *Building Trust in Information*, Springer Proceedings in Business and Economics, DOI 10.1007/978-3-319-40226-0_1

more recent times, however, many other disciplines have taken a keen interest in provenance and have begun to conduct research actively on what it is and how it can be effectively represented in different contexts. What has prompted this rising interest in the concept of provenance?

Without wishing to be overly deterministic, it is fair to say that increasing use of information and communication technology has been a major factor. The need to manage, preserve and make accessible new digital forms of recorded communication has caused archivists and librarians to look again at the principle of provenance, and how they apply it. In the field of law, provenance has become an ever more crucial aspect of the admissibility and weight to be given to electronically stored information as evidence. Computer scientists recognize a requirement to develop applications that trace and analyze the provenance of data across increasingly distributed and networked computing environments [2],[1] and in visual analytics, an emerging field that combines human and machine intelligence, there is recognition that the need to trace provenance extends beyond computing environments and into the realm of human analysis and reasoning.

This chapter presents a synthesis of the discussion that emerged from a workshop on provenance that brought together participants from multiple fields[2]: archival science, law, library and information science, computer science, and visual analytics. Each of these fields understands the meaning and purpose of provenance in subtly different ways. Participants of the workshop shared multidisciplinary and interdisciplinary perspectives on provenance and its application areas, with particular reference to trust in distributed computing environments. The aim was to create cross-disciplinary bridges of understanding with a view to arriving at a deeper and clearer perspective on the different facets of provenance and how traditional definitions and applications may be enriched and expanded via a multidisciplinary and interdisciplinary synthesis.

Why the need to look at provenance from a multidisciplinary and interdisciplinary perspective now? Many cognate disciplines are working on the challenge of representing provenance information in the context of digital recorded communications, but are doing so without reference to, and even in ignorance of, one another's approaches. Cross-fertilization between fields working on representation of "record" provenance (i.e., archival science) and those working on "data" provenance (i.e., computer science), for example, in some cases using the same standards and technology (e.g., semantic web specifications such as RDF), enrich the efforts of each field. The rise of multi- and interdisciplinary research also means that there is a need for a generic framework that can encompass multiple content types and use cases. There are also future developments to consider—such as 'mixed initiative'

[1] More specifically, Moreau's analysis revealed a growing trend of research activity related to provenance, with about half the papers concerning provenance published since 2008. He conjectured that the development of the Grid as a technology for running scientific applications and the UK e-science program have been two significant external triggering factors that have caused increasing numbers of researchers to focus on the provenance problem.

[2] Appendix A of this volume provides a complete list of workshop participants.

systems[3]—adding a layer of complexity and new challenges that are more easily addressed from an multi- and interdisciplinary perspective.

2 The Role of Provenance

Provenance, and its representation, has many different and interrelated uses. The role of provenance in the determination of trust, a concept that in itself is multifaceted and difficult to define, is however one of the main purposes identified across the disciplines.

In archival science, trusting an archival object (i.e., a record as evidence of the documentary facts contained therein) has to do with the belief that such objects are authentic and can be relied upon. Such reliance is often the result of a risk assessment—conducted either intentionally or not—wherein the significant properties of the object itself are analyzed and assessed [3]. Provenance—in the sense of the origination and custody of the archival object—is one of the key inputs into this evaluation. Historically, the placement of archival objects in trusted archival repositories (e.g., the Roman Tabularium) in the care of special custodians who carefully controlled access, established and maintained authenticity and reliability of archival objects as evidence [4]. Increasing digitalization, especially in the context of distributed and networked computing environments, has problematized the traditional archival approach to establishing and assessing the trustworthiness of archival objects, as many archival repositories continue to have no means of identifying, ingesting and preserving archival objects in digital form. Nevertheless, archival ideas about establishing the authenticity and maintaining reliability of archival objects continue to provide a valuable theoretical framework.

More recent establishment of trusted and secure digital repositories, such as the Bodleian Library's Digital Safe, is addressing the need to preserve digital objects. The provenance requirements of Digital Safe were considered by users to be paramount because of the highly sensitive nature of the data, such as patient records or closed digital archives containing personal data, private data or legally privileged information [5].

In law, a trusted chain of custody establishes trustworthiness, though less emphasis is placed on the curatorial aspects of trust in evidence. The Provenance of how a piece of evidence came into the hands of investigative authorities is a crucial aspect of the admissibility and weight to be given to such evidence [6]. Is this evidence the "best" evidence or a "copy" that may have differed from the original? How was the evidence produced? Was it generated in the normal course of business,

[3]Mixed-initiative systems can be described as systems that augment human cognitive capabilities by developing machines capable of offloading human thought processes and actively supporting individuals in pursuing their goals. In this sense, the focus is less on creating artificial intelligence (AI) than on augmenting human intelligence (IA).

or in response to a foreseen disclosure? Is it primary source material or secondary opinion? Is the evidence hearsay, or derived from personal knowledge of a person who has sworn an affidavit? Can that person be cross-examined? From where does this evidence derive? Who has had control over the evidence and the responsibility for its custody and control to maintain its integrity? Who has had the responsibility for the preservation and disclosure of the records that make up the evidence when demanded by the Court? Is there any additional evidence (e.g., metadata) that can be examined to ensure the evidence has not been altered while in the custody or control of the individuals or systems where the records have been maintained? Has there been an organized system of records management including a record archiving and destruction schedule that accommodates special preservation in situations of notification of potential litigation? These are all questions about provenance that are crucial to answer in the context of establishing whether evidence can be trusted, and therefore, admitted as evidence by a trier of fact in law [7].

Provenance as a means of assigning attribution is not only important in terms of establishing the degree of trust that can be placed in information, but also in terms of assigning rights, such as intellectual property rights. With the rise of open research, wherein organizations create and publish sets of open data that are generated and transformed through multiple autonomous information systems, and used, mixed and re-used by others, not only do issues arise about reliability and authenticity of such data, but also issues of credit, licensing and the right to benefit from exploitation of the data.

Another important purpose of provenance is its role in semantic interpretation. Every archivist knows that the feature that separates archives from other forms of information is that they derive their meaning and value from their provenance. If you do not know the provenance of a document, then the document is no more than a decontextualized source of information—an information object that is largely devoid of wider meaning and evidential value. Knowledge of the provenance of a document enables that document to be used as evidence of activities, for it is essential to know who created or received the document and for what purpose in order to interpret it. Knowledge of historical context allows the receiver of a communication from the past (i.e., the reader of an archival record) to continue to interpret its meaning in the present.

In knowledge organization, provenance plays an important role in understanding semantic changes to classification or ordering of ontological ‘things’ over time. Knowledge organization entails the grouping of specific instances of entities in a domain of analysis into semantically meaningful classes. These groupings frequently shift and change (e.g., the classification of Eugenics in the Dewey Decimal System [8]), and need to be tracked, for example, to support effective information retrieval.

Provenance also plays a role in information retrieval as one of the key access points in archives and libraries. A book, for example, may be retrieved by the name of it author. Bearman and Lytle [9] encouraged archivists to recognize provenance as a discovery tool to reveal the pertinence of records through information about the context of their creation and relationships among records. According to these authors, the effectiveness of provenance information as a means of archival discovery increases with its extent and standardization. The more that an archivist knows and conveys about the actors and actions involved with the creation and keeping of a body of records, the easier it will be for researchers to make use of the records for research purposes.

Somewhat relatedly, in visual analytics, provenance is seen as having a role in the process of sense-making. Wheat argues (this volume) that an external account of the history of the analysis is a key component of "reflection-in-action" [10]. Representations of provenance information and descriptions of how they have been used better enable an analyst, or a group of analysts working collaboratively over time, to interpret and make sense of heterogeneous information resources.

Transparency and accountability are two additional objectives of representing provenance information. Through the lens of computer science and various of its sub-disciplines, Moreau [2] has written: "A powerful argument for provenance is that it can help make systems transparent, so that it becomes possible to determine whether a particular use of information is appropriate under a set of rules. Such capability helps make systems and information accountable. To offer accountability, provenance itself must be authentic, and rely on security approaches." In the context of visual analysis, representations of provenance help those who may have to take decisions based on the visual analysis (e.g., policy makers), to trust the conclusions arising from it [11]. As Burgess (this volume) points out most descriptive metadata in libraries consists of unqualified assertions that are of limited utility to scholars and usually of somewhat variable quality. The addition of provenance information renders sources transparent, holds the librarian accountable for the reliability of these assertions, and encourages scholars to correct and enhance metadata (subject to requisite permissions) since provenance information can be attributed to a scholarly source. In the context of governance, access to information about the provenance of information ensures that government records and data can be trusted and used to hold public officials accountable.

Archivists also use provenance information to assess the value of archival records and make appraisal decisions leading to decisions about what to preserve and what to destroy [12, 13]. Archival macro-appraisal theory, for example, ascribes value in determining what to keep and what to destroy not on the basis of trends in historical research or predicted research uses but on the basis of how archival records provide evidence of the functions of the state and citizens interaction with these functions [14]. This shifts the primary focus of appraisal from the record—including any research characteristics or values it may contain—to the functional context (i.e., as part of its provenance) in which the record is created [15].

3 Defining and Conceptualizing Provenance

A careful reading of the above discussion will have revealed that there are important differences in the conceptualization of provenance among disciplines and communities of practice. In fields, such as those brought together for the workshop, where provenance is an active area of exploration and research, it is not surprising that the concept is dynamic and constantly being reformulated even within disciplines or cognate areas. This section presents an overview of what participants from distinct disciplines—archival science, computer science, library and information science, and cognitive science (as one of the disciplines underpinning the field of visual analytics)—shared about how provenance is conceptualized and defined in their areas and, in some, cases how the concept has evolved from its traditional usage to the present day, as a reflection of the dynamism and multi-dimensionality of the concept of provenance.

3.1 Archival Science

In archival science, provenance is often defined in terms of the organization or individual that created, accumulated and/or maintained and used records in the conduct of business prior to their transfer to a records center or archives. The *General International Standard Archival Description* (ISAD (G)) [16] and the *International Standard Archival Authority Record for Corporate Bodies, Persons and Families* (ISAAR (CPF)) [17] define provenance as the relationship (in ISAD (G)) or relationships (in ISAAR (CPF)) "between records and the organizations or individuals that created, accumulated and/or maintained and used them in the conduct of personal or corporate activity." In 2012 the Committee on Best Practices and Standards of the International Council on Archives (ICA) recommended adoption of the following definition: "Provenance. The relationship between records and the organizations or individuals that created, accumulated and/or maintained and used them in the conduct of personal or corporate activity. Provenance is also the relationship between records and the functions which generated the need of the records" [18]. The archival principle of provenance (sometimes referred to as *Respect des Fonds*) establishes that the records of a creator should be kept together, AND (by some accounts) should be organized according to the purpose and functions of the creator (sometimes referred to separately as the Principle of Original Order in English, *Respect pour l'Ordre Originale* in French, *Registraturprinzip* in German. and which also relates to the concept of archival bond) [19–21]. Thus, the principle of provenance in archival science usually applies to an aggregation of records (e.g., a *fonds* and/or a series), while in computer science (see below) the focus is on individual items. Cook [22], however, has said it can apply to individual records and archival documents or a group or series of archives.

These definitions misleadingly suggest that the archival concept of provenance is a relatively stable construct and tend to hide the considerable dynamism and debate that has taken place about the meaning and application of the concept almost since the inception of its use. Indeed, Ken Thibodeau (this volume) refers to it as "narrow", "vague" and "arbitrary," and Cunningham (this volume) suggests that it is poorly understood. At the First Stockholm Conference on Archival Theory and the Principle of Provenance, which took place in September 1993 [23] the speakers debated whether provenance was a universal scientific theory or mere concept for organizing archives; whether provenance included the internal arrangement of records within series or parts of the fonds as well as linking the entire fonds to its creator; whether provenance only underpins the evidential and primary value of records, but not their informational and secondary value [23].

Douglas [24] has observed that "much of modern archivists' criticism and discussion of the principle of provenance has focused on how to effectively represent the fluid and changing nature of both the external and internal structure of provenance of archival aggregations." Though not abandoning the traditional archival constructs, Michel Duchein, in his 1983 article "Theoretical Principles and Practical Problems of Respect des fonds in Archival Science," [19] noted that the practical application of the principle gave rise to many theoretical difficulties (e.g., how to ascribe provenance in cases when records creators contributed collectively to the creation of records in a shared database system). As early as the 1950s, the Australian archivist Peter J. Scott advanced the notion of what came to be called the "series system" as a reaction to the limitations he found in traditional approaches to administering archives according to the principle of provenance [25]. Scott's series system advocated the adoption of the series as the primary locus of intellectual control and description and the use of authority records to link series with as many records creators as warranted.

A fundamental shift in archival thinking came in the 1980s with rising use of technology and the introduction of standardized approaches to archival description. At the time that the *Canadian Rules for Archival Description* [26] were under development, Debra Barr and Terry Cook emphasized that an archival fonds should be viewed as an abstraction rather than as a physical entity, signaling the end of using provenance as a determinant of the physical ordering of records as it had been up to that point in much of the archival world, though it continued to be used as a unit of intellectual control [22, 27]. Bearman's and Lytle's [9] article on the power of provenance further cemented the shift away from viewing provenance as a determinant of physical arrangement and towards a view of its value as a tool for information retrieval. At a time when most archival descriptive systems only accommodated simple hierarchical orderings (i.e., as in each body of archival records was affiliated with one function/creator in a one-to-one relationship), coming out of a tradition of museum informatics, Bearman was already advocating an archival descriptive system that expressed one-to-many relationships.

In the 1980s, Barr criticized the conceptualization of the fonds proposed for Canadian rules of archival description as being too reductionist, stating that "Respecting provenance means reflecting more than one aspect of the complex history of many records" [27]. Notions of provenance continued to expand as a means of better representing and preserving complex recordkeeping realities. In his 1992 article [22], "Mind over Matter", Terry Cook posited that the fonds is created through description of relationships (e.g., between the program (function), the agency (structure), and the citizen.) and that provenance lay "at the heart" of these relationships.[4] Millar [28] has argued for a notion of provenance as encompassing creator history, records history, and custodial history. Nesmith [29] also has been critical of the reductionism inherent in current forms of archival representation of provenance. In his paper "Reopening Archives: Bringing New Contextualities into Archival Theory and Practice," [30] Nesmith calls for archival description to be thought of as "the action mediated by archivists of researching and representing the multi-faceted contextuality (or history of records or 'archival narrative' about them) which enables records and knowledge to be made through archiving."

Lemieux [31] and Ken Thibodeau (this volume), influenced in part by theories deriving from mathematics and computer science, argue that provenance can be expressed as a mathematical graph. Thibodeau sees records as the nodes and the edges as the links between records that result from the same activity, representing the archival bond. Lemieux [31], influenced by Cook's conceptualization of provenance as a network of relationships e.g., among records, programme and agency [22], argues that that a provenance graph should incorporate archival records as nodes as well as differentiated edges to represent the different relationships (e.g., of records to agents, agents to functions, records to functions, etc.).

3.2 Library and Information Science

In library and information science and, specifically, knowledge organization, the literature discusses versioning, instantiation, and ontogeny.[5] It does not mention provenance, though Tennis (this volume) argues that these are all related concepts. Tennis' work focuses on provenance as defined as the chronology of custody and location of (library) material, and how revisions of indexing languages could change the location of a concept. With the change in location, the concept may change its meaning, and it is the meaning of the concept, in relation to other concepts and the documents they index that is the focus of knowledge organization [8]. These notions

[4]There is some debate about whether Terry Cook was referring both to creators (agents) as well as to function in his reference to relationships. In 2010, in his International Council on Archives plenary address, Terry Cook stated his view of the concept of provenance: "provenance is the concept of linking records or archives, or group or series of archives, to their creator, whether an individual or organization. The value of provenance is that it allows archivists and researchers to understand a record and its content in terms of who made it, where, when, how, and why, and what changes have taken place with the record over time, and why" and from this Duranti infers: In short, Cook appeared to equate the term "provenance" with "context", but his "who made it, where, when, how, and why" definitely refer to persons, not functions. Lemieux (2014), on the other hand, finds evidence that Cook was also referring to functions. She cites his 1992 article "Mind over Matter", in which he writes in reference to his provenance-based macro-appraisal theory, "Turning to the second part of the model, the citizen-state interaction reflects a convergence of three factors: the programme (function), the agency (structure), and the citizen."

[5]When we trace the history of a concept through revisions of indexing languages, we are studying the concept's ontogeny. Ontogeny is a term borrowed from biology.

have particular resonance with ontology-based approaches to data management, which are taking root in financial information management with the implementation of the Financial Industry Business Ontology [32].

3.3 Computer Science

Moving to computer science, specifically in the context of linked open data on the Web, the meaning of provenance has been codified by W3C PROV-DICTIONARY [33] as: "a record that describes the people, institutions, entities, and activities involved in producing, influencing, or delivering a piece of data or a thing." This definition reflects the fact that, from a computer science perspective, "the goal is to conceive a computer-based representation of provenance that permits useful analysis and reasoning ..." [2].

As in archival science, in computer science the conceptualization and definition of provenance demonstrates variety and evolution. Luc Moreau's (2010) paper [2], which offers a comprehensive survey of the technical literature on provenance, illustrates this point. Through a survey of multiple sources, including the author's own original database, the ACM, IEEE, and Springer digital libraries, the DBLP computer science bibliography, and some programmes of provenance-specific events such as the International Provenance and Annotations workshops (IPAW'06, IPAW'08), and the Workshop on Theory and Practice of Provenance (TAPP'09), he identified a total of 425 papers, with the first publication dating from 1986 and describing an auditing technique to assist analysts in understanding and validating data results. Moreau also observed several publication peaks coinciding with events organised by the "provenance community" as he defines it: in 2002, the first provenance workshop organised by Foster and Buneman; in 2006, the International Provenance and Annotation Workshop (IPAW) workshop organised by Foster and Moreau; and, in 2008, the second IPAW workshop organised by Freire and Moreau and the first Provenance Challenge special issue edited by Moreau and Ludäscher [2], noting how conceptualization and definition of provenance was shifting with development of the Grid as a technology for running scientific applications and the UK e-science program. Based on his analysis of the literature, Moreau [2] extrapolates a definition of provenance as: "(Provenance as Process) The provenance of a piece of data is the process that led to that piece of data," noting that many things pertaining to execution may be captured under "process", including the executed program, input data, configuration, computer, electricity powering it, users, etc. Moreau further notes several other definitions of provenance, including provenance as a directed acyclic graph, which, interestingly, illustrates the influence of computational approaches on archival thinking and a convergence between computer science and archival science approaches to provenance (cf. [31]). However, this definition is arguably less definition than a way of expressing and representing provenance. Moreau further identifies "Provenance as Annotations" such as Dublin Core metadata standard [34], designed to provide structure and semantics to metadata of resources, noting that aspects of these ontologies are provenance related, such as author, creation date, and version and arguing that such information can also be seen as a specialization of a process-oriented definition of provenance. Finally, he identifies an event-oriented conceptualization of provenance, according to which provenance is a chain defined as a time-ordered sequence of provenance records capturing events affecting a document [2].

At the same time, the theory of data provenance progressed on a parallel track. The work of Buneman et al. [35] on provenance, in the context of databases, was a milestone for that research. They defined data provenance, which they note is sometimes also called "pedigree" or "lineage", as the description of the origins of a piece of data and the process by which it arrived in a database. They further drew a distinction between "where" provenance (e.g., where does a given piece of data come from) and "why" provenance (e.g., why is it this database). Within data management and information processing other formal definitions have been given for data provenance. The provenance of a data item that is returned by a database query, for example, encodes the semantics of the query itself, and mentions the fragments of the database state that were involved in the query processing [36]. An algebraic theory in support of data provenance representation and management has been developed [37]. This form of fine-grained provenance is often contrasted with coarse-grained provenance, which records the input/output derivations that are observed when functions are invoked, typically from within workflows and in the context of scientific data processing [38]. Attempts have also been made to reconcile these two views, e.g. when declarative-style queries are embedded within procedural workflow processing [39].

3.4 Cognitive Science in Visual Analytics

Visual Analytics, "the science of analytical reasoning facilitated by interactive visual interfaces" [40] is yet another discipline—or, rather, interdiscipline, being comprised of several other disciplines including computer science, cognitive science, and visual art and design)—evidencing growing research interest in provenance. Given the relative newness of visual analytics (established c. 2005), conceptualizations of provenance have not yet been codified into formal definitions, but rather must be extracted directly from the literature. Jankun-Kelly [41] defines provenance as: "information about entities, activities, and people involved in producing a piece of data or a thing, which can be used to form assessment about its quality, reliability or trustworthiness." This very closely resembles the W3C PROV definition mentioned above, perhaps reflecting the close affinity that exists between visual analytics and computer science.

Distinct from computer science, however, the notion of analytic and/or reasoning provenance has emerged from within visual analytics. For instance, Keim et al. [42] describe reasoning provenance as documenting the entire analytic process, including provenance data and details of findings and discoveries. Xu [43] further highlights this distinction, positing, "Different from *Data Provenance*, which captures the information about data collection and (computational) processing, Analytic Provenance describes the human (such as an analyst) data exploration and reasoning process. Examples include how an analyst uses a Visual Analytics system to interactively explore a large dataset (data exploration), and how he/she pieces together findings to form hypotheses and seeks evidence to validate or reject them (reasoning)." External representation of provenance information is a key component in support of the sense-making process [44, 45] in the context of distributed cognition environments [46, 47]. According to Xu [43], analytic provenance can be used to support sense-making in the following ways:

- Support to the individual analyst, as a resource for "reflection-in-action" [10]. Analysts can use the analytic provenance information (through visualization) to understand the sense-making space, review progress, and plan and manage further analysis.
- Support to sense-making communication in collaboration and reporting. By capturing and visualizing analytic provenance, analysts can share with team members not only the findings but also the sense-making processes that led to them, allowing for constructive critique and important for handover. Similarly, it enables reporting to decision makers not only of the conclusions but also the reasoning processes that led to conclusions, allowing for judgment of confidence.
- Tracking data quality and human bias. One of the main focuses of analytic provenance research is to capture and model data quality and its impact on sense-making. This entails determining how data quality issues are (visually) represented and how this information is used in sense-making. Analytic provenance information can help detect and track human errors (such as confirmation bias) and their impact on the sense-making findings [43].

Synthesizing various perspectives, Roberts et al. [48] suggests that provenance in the context of visual analytics can be examined at a number of levels:

- At a data level, taking into account that all data will have some source, and a path between this source and its use in analysis.
- At the analysis level, accounting for the actions performed and techniques used in the analysis at a given point.
- At the reasoning level, dealing with the way the conclusions in analysis has been reached.

4 Domains of Application and Use Cases

There exist a wide range of contexts or domains, herein represented as use cases, in which the application of provenance is needed to meet prescribed objectives, such as the ones identified in the section on the role of provenance. This section offers a brief overview of the main use cases referenced by workshop participants.

A common thread amongst all of the use cases is the complexity of conceptualizing and representing provenance information from heterogeneous sources across increasingly complex and interconnected systems.

4.1 Preservation of Digital Records and Other Digital Materials

Of concern to a number of the participants (i.e., professional archivists and librarians) was the necessity of representing provenance information in order to preserve digital objects, whether these objects are archival records, books, manuscripts, or some other digital artifact.

The Bodleian Library, for example, collects research outputs of the University (journal articles, book chapters, PhD theses, conference proceedings, and increasingly, datasets underpinning publications), as well as the products of in-house digitization of library material and born-digital archival deposits. Capturing, preserving and disseminating provenance information about such a huge array of materials presents an enormous challenge [5]. Preservation takes place over a time scale during which technologies, formats and preserving communities are very likely to change. Thus, specialized approaches, models and technologies are needed to guarantee the long-term understandability of the preserved data.

Records, which have traditionally been the purview of archivists to preserve, are no longer made as simple, static documents in the form of text on the page or pieces of paper; they are not even made of simple digital documents any longer in many cases. Instead, there are many new forms of complex and interactive digital objects, such as linked open data, and even interactive visualizations, that require solid information on provenance to ensure their reliability and authenticity as records. Factor et al. [49] argue that maintaining the authenticity (trustworthiness) and provenance (history of creation, ownership, accesses and changes) of the preserved objects for the long term is of great importance, since users must be confident that the objects in the changed environment are authentic and reliable, and call for new "preservation-aware" systems.

4.2 Cloud-Based Storage

Many organizations are increasingly looking to cloud-based technology as a place to store digital objects. In a cloud environment new challenges arise. For instance, in order to manage client data, a cloud service provider (CSP) takes a certain level of control over that material. When records or data are entrusted to cloud systems, creator-generated metadata are also stored, and CSPs assume control of the material. Within this new environment, these user records will acquire additional metadata from the CSP that will be indicative of a number of important elements, including, but not limited to, storage locations, access controls, security or protection measures, failed or successful manipulations or breaches, etc. CSPs may also outsource some components of their services to other third parties, who may also generate service metadata that provide assertions about the maintenance and handling of the material, and about their own actions taken in the course of handling the material. While these metadata are linked to users' records, much of it remains proprietary to the provider and not the user. Consequently, proprietary CSP metadata present a sort of event horizon, beyond which the ability to establish an unbroken chain of custody is lost to the owner of the records. CSPs remain reluctant to share information about the cloud environment itself, the movements of a client's data within the system, and when the provider (or its contracted third parties) might have access to the data. Additionally, the network of third-party subcontractors employed by a provider may make it impossible for them to know such information. Nevertheless, these metadata remain invaluable to the user in assessing and ensuring the accuracy, reliability, and integrity of the material over the whole service lifecycle [50]. Is there a way in which a balance might be struck between a provider's desire to protect the confidentiality of their business processes and trade secrets, and a client's need to ensure trustworthy records in the cloud? Much of the reluctance to engage cloud services might be mitigated by transparent and standardized metadata that is collected, managed, and then shared with users by CSPs [50, 51]. An InterPARES research team [49] investigating digital preservation in the Cloud is designing a model and a set of functional requirements for preservation of digital records, in order to provide insight and guidance to both those who entrust records to the Internet and those who provide Internet services for records.

4.3 Digital Evidence in Litigation

The Sedona conference guidelines [52] on e-discovery begin with the observation that "Today most information created and received in organizations of all sizes is generated electronically in the form of e-mail messages and their attachments, word processing or spreadsheet documents, webpages, databases and the like. Even formal documents—such as tax returns, applications for permits and other documents filed with regulatory authorities—generally originate and often are filed in electronic format. Much of the information is never reduced to paper. Meanwhile, because of how computers operate, vast amounts of electronic data are created and maintained—seemingly forever–often without users even knowing that the data has been created, much less saved. Yet while this data is kept 'seemingly forever,' due to changes in technology, it may rapidly become inaccessible unless migrated to new formats."

4.4 E-Science and Reproducible Research

There is growing recognition in the sciences that journal articles are insufficient when it comes to enabling the reproducibility of research, one of the cornerstones of the scientific method. This awareness led initially to requirements for mechanisms to enable data publication, dissemination and discovery but subsequently, also to the realization that data was of limited utility without methodological metadata (couched in very similar terms to historical provenance) alongside. The reuse and reproduction of scientific experiments as they are described in publications can be hard. Often it requires additional information, data, tooling or support beyond that provided in the text of a traditional publication. As an example of work being done in this area, the NSF DataONE[6] project is currently the largest Research Data conservancy project in the USA, with a focus on Earth Observational Data. Metadata is the foundation of effective search capabilities across a large collection of Science Data Objects (around 180,000), and provenance is an essential part of it [53]. The gathering of provenance information is, furthermore, an important approach that scientists use to gain confidence in their conclusions. Data come from heterogeneous sources: as part of one research investigation there may exist, for example, slides hosted on slideshare; code in a Github repository; data in Figshare; and data in ArrayExpress. A growing number of activities are developing new mechanisms, or repurposing existing mechanisms in order to describe and associate resources like this together, in a machine-readable manner, so that they can be more easily shared, and exchanged. The goal is to improve the potential for understanding and reuse of research by making sure that the information that is needed to make a published resource useful is associated with it, and shared.

4.5 Digital Humanities Research

Similar to developments in e-science, new digital forms of humanities research require management and preservation of diverse resources across collaborative research platforms. An example of this is the Oxford-based Cultures of Knowledge Project,[7] which, since 2009, has been using a variety of research methods to reassemble and understand early modern networks of communication. As it moves into its third phase (April 2015–March 2017), project participants are aiming to create a central repository of sixteenth-, seventeenth-, and eighteenth-century correspondence populated with metadata drawn from the widest variety of sources worldwide, and increasingly representative of the early modern "Republic of Letters" as a whole (see, Burgess, this volume). They are pursuing this aim by

[6] See https://www.dataone.org.

[7] See http://www.culturesofknowledge.org.

experimenting with a variety of methods and approaches to metadata aggregation simultaneously:

- Ingesting the *digital catalogues of major scholarly projects* in this field and linking through to their digital archives where available;
- Ingesting the *digital catalogues of collections and archives* with rich holdings of relevant material;
- Collecting the *digital files of recent and forthcoming printed editions and inventories of correspondence*, from which metadata can be extracted efficiently and accurately;
- Scanning *existing printed inventories of correspondence* and outsourcing their keying;
- Piloting *controlled crowd-sourcing* of metadata for key correspondences via a distributed community;
- Publishing *digital images* of corpora of learned correspondence and inviting collaborators to catalogue these letters directly within the editorial interface.

The volume of material, in digital and print form, potentially included in the catalogue by combining these methods is vast and the task of documenting its provenance and ensuring its integrity complex.

4.6 Open Data

An unprecedented number of individuals and organizations are finding ways to explore, interpret and use Open Data, defined as data that can be freely used, reused and redistributed by anyone subject only specific licensing requirements. An example is Open Research data reuse [54]. Public agencies are hosting Open Data events such as meetups, hackathons and data dives to exploit the possibilities inherent in openly available data. The potential of these initiatives is great, including support for economic development [55], anti-corruption [56] and accountability [57]. However, the quality, and in particular, the integrity, of open data is problematic. A recent UK report notes that poor data quality is hindering the UK government's Open Data program [58]. Far from being a one-off problem, research suggests that this issue is ubiquitous and endemic. Some estimates indicate that as much as 80 % of the time and cost of an analytics project is attributable to the need to clean up 'dirty data' [59]. As part of the wider data quality issues, data provenance can be difficult to determine. Knowing where data originates and by what means it has been disclosed is key to being able to trust data. If end users do not trust data, they are unlikely to believe they can rely upon the information to serve the purposes for which they are using it. Further, full comprehension of data relies on the ability to trace its origins. Without knowledge of data provenance, it can be difficult to interpret the meaning of terms, acronyms and measures that data creators may have taken for granted, but which are much more difficult to decipher over time. Establishing data provenance entails a good deal of effort undertaking activities including enriching data with metadata such as the date of creation, the creator of the data, who has had access to the data over time and ensuring that both data and metadata remain unalterable.

4.7 Crowd-Sourced Knowledge Management Platforms

Like the growth of Open Data, there has been growth in the development and use of crowd-based knowledge production in many domains. Perhaps the best example of this is Wikipedia, but many research libraries and archives are also turning to crowd sourcing of descriptive metadata for their collections (see, for example, [60–62]). Again, trust, related to the integrity of the entries on the site is a key issue. For instance, Wikidata acts as central storage for the structured data of its Wikimedia sister projects including Wikipedia, Wikivoyage, Wikisource, and others, and also provides support to many other sites and services that rely upon its data as input into their applications.[8] Errors and other problems in upstream Wikidata have the potential to cascade through the data supply chain. Traceability of data, through provenance information, is essential to establishing the authenticity and reliability of such data and can have a major impact on social participatory models.

4.8 Knowledge Organization and Indexing

Indexing languages are tools used in the aid of information retrieval and sense-making. They comprise classification schemes, thesauri, ontologies, and taxonomies. As more literature is added to the collection represented, and as users' needs change, so too do indexing languages. This causes a shift in structure and semantics in the indexing language. For example, in the 1913 Dewey Decimal Classification (DDC), number 397 was the single address for Gypsies, Nomads, and Outcast Races defined as: "[p]eople without nationalities who do not coalesce with the ruling people among whom they live. This includes Gypsy language, which has no place in the linguistic groups of 400, as the Gypsy people have no place in the geographic divisions of history,"[9] [63]. This phenomenon, while rich with examples from Dewey because of its age, is not the only indexing language that changes. The Wikipedia category system is another example. Because changes in indexing languages is a persistent phenomenon, and there is no commonly accepted design amelioration, it constitutes an important research area in knowledge organization [8].

---

[8] The content of Wikidata is available under a free license, exported using standard formats, and can be interlinked to other open data sets on the linked data web. (http://www.wikidata.org/wiki/Wikidata:Main_Page).

[9] 397 is between 396 Women's Treatment and Position and 398 Folklore, Proverbs (s.l.).

4.9 Intelligence Analysis

In intelligence analysis, the sources of data may not be known. Further, data comes with associated reliability issues, such as source uncertainty (see, Varga, this volume). It is therefore crucial to know the reliance that can be placed upon data. The capture of analytical provenance: the actions taken to perform an analysis within a visual analytic system can be captured: events (e.g. a key press) or actions (e.g. a zoom) can be logged easily. The overall history of interactions can be recorded. However, it is much harder to capture an analyst's reasoning process. Although this externalization can be achieved through think-aloud protocols [64], the process can potentially change the nature of the reasoning and may reduce task performance. Moreover, analysis of such externally captured data is extremely time consuming. Nevertheless, traceability of the provenance of analytic conclusions (e.g., what information resources were relied upon? How reliable were they? How were they visualized? How did the analyst interact with the visualization? What interactions and visual displays prompted particular insights? What conclusions did the analyst draw from these?) is extremely important given the impact of decisions based on such conclusions.

In intelligence analysis, there may be a further requirement for non-disclosure of provenance information in some settings where security concerns require that disclosure of provenance be limited to certain parties. A typical example is found in the context of intelligence sharing, where the provenance of the intelligence may contain insight into the process that led to the intelligence being collected. As that is often privileged information, models and mechanisms are needed to abstract out parts of it, depending on the clearance levels of the recipients.

4.10 Decision-Support Systems

Decision-support systems such as those for disaster recovery or systemic financial risk analysis rely on a wide range of data inputs from different sources. In addition, they usually involve heavily regulated domains with specific guidelines: international, national, regional and site-specific rules govern how decisions are made. Application of rules must be ensured, be auditable and may change over time. For example, the US Federal Open Market Committee meets every 5–6 weeks to implement the Federal Reserve's monetary policy [65] by deciding on open market operations. Formal decision processes typically generate agendas and minutes for the public record. Decision-makers often receive formal or informal advance briefings; the "arbitrary and capricious" standard for accountability [66] implies a need for solid analysis and strong documentation. One benefit is the reduction in uncertainty from a conversion of complex, subjective, and ambiguous information into a clear ruling. The process of introducing or modifying regulations is highly formalized and open to public scrutiny, which is often extensive. Careful documentation of provenance allows tracking the sequence of decisions steps, which is crucial in maximizing the efficiency and recovery rate from a disaster or other risk scenario [65].

4.11 "Human-in-the-Loop" Processes

Cognitive systems encompass a range of technologies, including artificial intelligence, expert systems, and human-machine interaction interfaces, and encompass systems that are capable of learning from their interactions with data and humans [67]. These systems afford mastery of the tasks on which they work, as well as extraction of contextual information from the environment in which these tasks are undertaken. They solve problems as they arise and plan for the future. They communicate appropriately with others about themselves and their activities in order to work effectively in close collaboration. They also adapt their understanding and skills as they and the world around them evolve [68]. Traditional computers are organized around microprocessors. With cognitive systems, it is much more about the data and drawing insights from it through analytics [67]. Though the ultimate goal may be to create intelligent machines, the next logical step in the evolution of cognitive systems is to augment human cognitive capabilities by developing machines capable of offloading human thought processes and actively supporting individuals in pursuing their goals. Such cognitive systems serve the individual by reducing their cognitive supervisory burden. "They enhance the individual's cognitive abilities by supplementing memory and problem-solving capabilities and by providing direct access to relevant data, expertise, guidance, and instruction. They work towards shared goals while understanding enough about the task, the individual, and each other to assist, mentor, cooperate, and monitor as needed. And they reduce the individual's performance degradation by offloading activities and by anticipating the kinds of errors that tend to occur in stressful situations" [68]. Such "human-in-the-loop" processes require the capture of human interactions with information systems through a user interface.

5 Methods of Capturing and Representing Provenance

Just as each of the disciplines or communities of practice participating in the workshop conceptualized and defined provenance in different ways, each of these fields also have diverse methods of representing provenance. This section presents how workshop participants described the different methods they use to capture and represent provenance information. The approaches range from those that are highly technical in nature to those that do not use technology, and from methods that capture and represent provenance information at the point of creation to others that involve forensic analysis of previously captured provenance information or combinations of both approaches. Participants also discussed choices made and challenges faced in relation to capturing and representing provenance information within their fields. In some cases, they also discussed the effects of technology (e.g., online or distributed environments) on practices surrounding capturing and representing provenance information.

5.1 Law

In law, an assessment must be made of electronically stored information (ESI) submitted as evidence. In this context, the provenance of the ESI is a *fait accompli*; the trier of fact must forensically assess its admissibility. This is sometimes done in *voir dires* (i.e., mini trials held as part of criminal proceedings or during litigation to determine the admissibility of proffered evidence). With the advent of social media and the digitization of documents, images and even sound recordings, *voir dires* concerning the provenance and chain of custody of ESI have become much more common as part of the discovery and disclosure process of litigation. During *voir dires*, the admissibility and weight of ESI is determined based upon answers to the questions about the provenance of evidence and other evidentiary concerns such as, probative value, prejudicial effect, relevance, authenticity, and weight. Often the process of evidence gathering and the responsibility for the care and custody of ESI make the provenance of how it came into the hands of investigative authorities a crucial aspect of the admissibility and weight to be given to evidence [69].

5.2 Archival Arrangement and Description

In archives, as previously mentioned, capturing, representing and preserving information ("respecting") provenance is a fundamental aspect of the preservation of archival records. Capture and representation of provenance information in the context of preservation of archival records has undergone major transformation over time. Traditionally, capture of archival provenance information has been a forensic process involving historical research. For instance, provenance may be captured from a diplomatic analysis of the materials to identify creators and other relevant agents. Then, any report, accession register, or finding aid may help in reconstructing the chain of custody of the materials. Direct witness from any agents (creators, managers, archivists, users) may also help in piecing together the provenance of archival materials. The biography of individuals or administrative history of organizations that created and/or managed the materials, along with their mandates and competences, may also help understanding. Physical characteristics of the materials may sometimes be of little help. In the digital environment, metadata associated with or embedded into materials provide some relevant information on provenance, as will information about systems in which materials have resided.

In brief, provenance can be captured—mainly manually—from all these sources. Other aspects of provenance, more typically associated with chain of custody, have been captured in such documents as accession files and registers as part of archival administrative processes. Until recently, this provenance information was not typically made available to the researcher, and is still not made available universally in archival descriptive systems as a matter of course.

In regard to representation, most commonly, provenance is expressed textually in archival descriptions or as an attribute of a set of records in an archival control system. Such representation is usually carried out nowadays adopting some archival standards (e.g., ISAD(G) [16]); however, this was not always the case and, hence, archival descriptions exhibit a wide variance in quality and content (often much to the bemusement and frustration of researchers!). As an example, for reasons discussed in the previous section on conceptualizing and defining provenance, the Australian system of archival description (discussed below) differs significantly from European and North American systems. The main national and international archival standards (discussed in the following section) have some specific information elements conveying information on provenance, though such information may be dispersed throughout different metadata elements. Provenance information is not always explicitly identified as such; for example, both the printed National Archives Guide and the Online Public Access system of NARA capture provenance under the heading, "Creator."[10] This might be supplemented by additional data, such as successor organizations, if any. Some archival scholars are critical of these models and believe they do not represent provenance adequately (see, for example, [70]).

Australian archivists have evolved a unique approach to representing the provenance of archival documents. The Australian approach to representing provenance, based on using the series as the primary unit of intellectual control, consists of two inter-related component parts:

- *Context Control*, which is achieved by the identification and registration of records creating and other ambient entities and the documentation of the administrative and biographical histories of those entities, their functional responsibilities and their relationships with each other and with the recordkeeping systems they maintain(ed); and
- *Records Control*, which is achieved by the identification, registration and documentation of record series and/or the items that make up those series.

Within series systems implementations instances of each of the three main entities may be described at different levels of granularity, with relationships between the different levels described accordingly (Fig. 1) (Cunningham, 2015).

In the Australian system the contextual entities that need to be documented and linked to descriptions of records include individuals, families, organisations, project teams, government agencies and portfolios, governments themselves, functions and

[10]NARA. Online Public Access. http://www.archives.gov/research/search/.

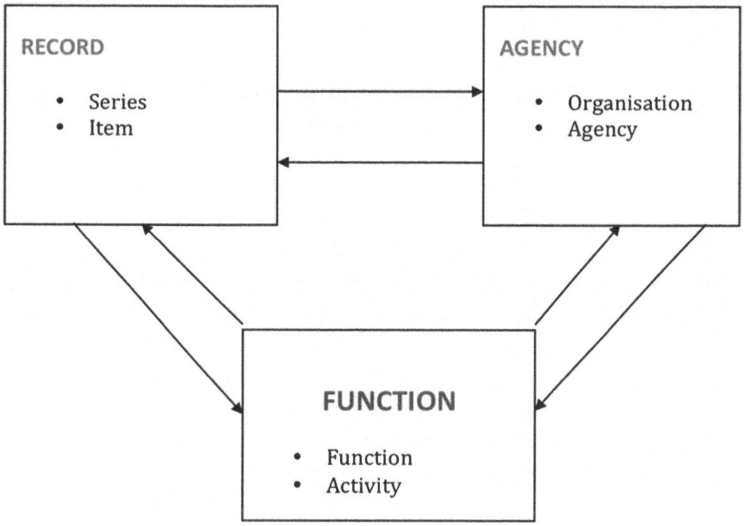

Fig. 1 High-level data model for provenance information captured in Australian archival control system

activities. As Cunningham (2015) points out, it is the complex web of dynamic relationships between these various entities that underpin the transactions that cause the creation of records. It is therefore essential to capture documentation of these relationships in order to provide the contextual knowledge necessary to understand the content of the records themselves. The Australian approach allows for considerable dynamism: structural elements of the system provide the conceptual and documentary building blocks from which traditional or non-traditional finding aids can be constructed as and when required (see, Cunningham, this volume). Furthermore, unlike traditional post-hoc approaches to archival description that focus on the static description of non-current records, the Australian approach is also used to capture and represent provenance in an active recordkeeping domain. This follows an approach first put forth by Bearman [71] who said, "archivists should find, not make, the information in their descriptive systems," implying that they should reuse and add contextual value to the metadata dynamically created in the records systems of records creators.

There is growing call for adoption of RDF in representing provenance information in archival arrangement and description in order to more fully express the complex relationships between records and the context of their creation and subsequent preservation (see, for example, [31]). However, this poses a new challenge, since provenance is represented through standards and models (e.g., the PROV-O Ontology) that are not specific to the archival domain and therefore may not be fully consistent with archival preservation needs.

5.3 E-Science

Somewhat similar to the traditional archival finding aid, in e-Science today the notion of a data paper has emerged. A data paper is a searchable metadata document, describing a particular dataset or a group of datasets, published in the form of a peer-reviewed article in a scholarly journal. Unlike a conventional research article, the primary purpose of a data paper is to describe data and the circumstances of their collection, rather than to report hypotheses and conclusions [72]. Like the archival finding aid, the data paper represents, and serves as a guide to, a complex recordkeeping reality comprised of many information resources discussed in the e-Science use case in the previous section and which institutions, such as the Bodleian Library and other repositories of scientific research data, must preserve.

In this context, provenance is captured through observation of a process in execution—a database query or a workflow—including processes carried out by humans or only partially automated. The nature of the process and the infrastructure onto which it is enacted determine the level of detail that is available to the observer. Provenance capture methods differ for each of these scenarios. Although an increasing number of tools and systems are being retrofitted with provenance recording capabilities, these are still few compared to the number of data processing environments used across disciplines. These systems include workflow management systems [38, 73], and more recently, the Python [74] and the R languages [75] for scientific data processing. The case of completely automated processes that run in a centralized environment is, however, the simplest possible scenario.

Human-in-the-loop scientific discovery processes are obviously more problematic, and are still, by and large, limited to capturing human interactions with information systems through a user interface. One of the open problems concerns capturing provenance from processes that are distributed over multiple, heterogeneous, autonomous systems. Each of these systems may be expected to provide some fragment of provenance, requiring post hoc composition of these fragments. A problem arises, however, when data identifiers are not used consistently (i.e., the same dataset if referenced in different provenance fragments, using different identifiers). There does not yet appear to be a systematic approach to deal with this problem, other than by issuing standard data identifiers (DOIs) and manually enforcing their consistent use in specific cases.

5.4 Digital Preservation in Libraries

As an example of how the capture and representation of provenance is implemented in a digital library context, the Bodleian Library *Digital Safe* captures contextual information about the digital objects stored in the repository as follows:

- (software, instruments, etc) that can create or change digital objects

- Locations—precise coordinates and geopolitical entities that define geographical scope for assertions
- Collections, works and instances—framing the more traditional library objects
- Events—the key provenance objects that embody the essential creation/modification/deletion of objects
- Annotations—additional assertions about relationships/properties of objects but also textual descriptions of provenance that have been inherited from various sources.

Provenance information is drawn from a wide range of sources given the nature of the collections, including online web forms, spreadsheets, TEI (for manuscripts) and EAD (for archival documents) encoded text records, and data feeds from repositories (OAI-PMH and other RESTful protocols) and databases (via machine-to-machine APIs) (see, Burgess, this volume). The RDF/lined-data model provides a generic mechanism for expressing this information [5]. Provenance is represented internally using W3C PROV-O in the context of the CAMELOT data model that underpins the institutional repository. The principal reason for using PROV-O was a requirement for an RDF description of relationships in the context of an event. The PROV-O data model does not, however, meet requirements for the representation of time, as it does not allow for incomplete time information (e.g., when only the year or decade or day is known). With a semantic data model it is possible to use more than one vocabulary and some entities defined in a data model and not others, as long as there is no conflict when using multiple vocabularies. Thus, it is possible to use PROV-O representation of an event, together with another vocabulary's representation of time. CIDOC-CRM is a vocabulary that has a good representation of approximate time for an event, arising, as it does, from representing activities relevant to museums [5] (and see also Burgess, this volume). There is a tendency for library metadata standards to become over-prescriptive—sacrificing fidelity to adherence to a data model. A key feature of linked-data RDF is that it permits extensible and flexible knowledge models. There is a careful balancing act required between standardization and fidelity to ensure interoperability between systems while providing the flexibility to express and capture the assertions that scholars wish to make.

5.5 *Knowledge Organization*

Indexing practices within the field of knowledge organization is an area where technology's effect can be observed. Tennis (this volume) notes that, in the online environment, there is a challenge with tracking changes insofar as those changes are related to a particular version of an indexing language. This is because instead of issuing particular editions, as was the case in the print-only world, it is now possible to change the state of an indexing language by changing a single term without creation of a new formal edition of the indexing language. Both states and editions constitute versions of an indexing language, and the provenance or ontogeny of a concept can be observed through these two kinds of changes to its instantiation. That is, the persistence or discontinuity of a concept is observed through observations of the indexing language, and specifically its terms.

5.6 Decision-Support Systems

In the case of decision support systems, Varga (this volume) following [76] notes that the provenance of the data used and decisions made are collected at their point of entry to the decision support system: the information on the data, who supplied the data, the time stamp and who made the decision (with the time stamp linked with the data available at the time the decision was made). Compliance and deviation from any recommended guidelines can also be recorded.

5.7 Visual Analytics

Visual analytics being a relatively new discipline has developed no standard approach to representing provenance and no single approach to its capture and representation in visual analytics systems. However, as an example, Gotz and Zhou [77] suggest that, as a basis for representing and capturing analytic provenance, it can be categorized using a four-layer hierarchical model based on its semantic richness. Figure 2 shows this model using an analysis of the stock market as an example: the level of semantics increases from bottom to top. The bottom-level *events* consist of low-level user interactions such as mouse clicks and keystrokes, which have little semantic meaning. The next level up is *actions*, which are

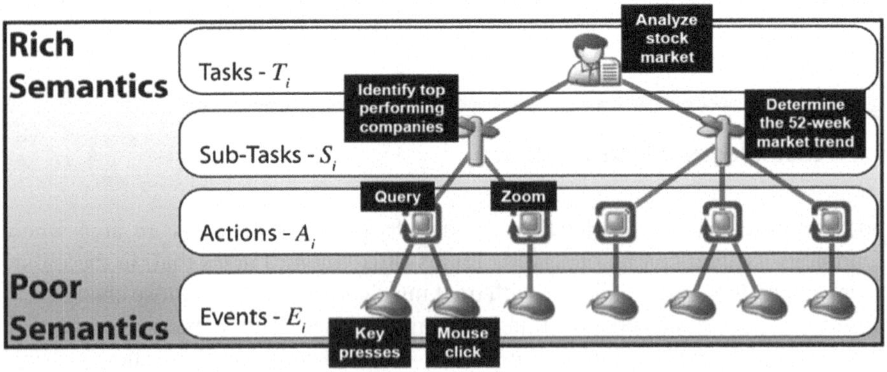

Fig. 2 The hierarchical analytic provenance model shown with an example of analyzing stock markets [77]

analytic steps such as querying the database or changing the zooming level of data visualization. The parameters such as data description and visualization settings are also part of the provenance. Further up are the *sub-tasks*, which are the analyses, required to achieve a sense-making goal. In the case of stock market analysis, examples are identifying top performing companies and determining long-term trends. In the top-level is the *task*, i.e., the overall sense-making undertaking, which is "analyzing stock market".

Provenance information (e.g., of the data used and analysis made) is collected at their point of entry to the decision system. Following Gotz and Zhou's four-layer model, capturing lower level events and actions is relatively straightforward in a visual analytics system. However, such analytic provenance information alone is of limited use [77]. Tasks and sub-tasks provide important clues to the purpose and rationale that underlies the sense-making. However, they are largely part of users' thinking, to which a visual analytics system does not have direct access.

Existing approaches to capturing high-level analytic provenance can be broadly categorized into *manual* and *automatic* methods. The manual methods [78–81] largely rely on users recording their analysis process and sense-making tasks, whereas the automatic methods try to infer the higher level tasks and sub-tasks from lower level events and actions. While the manual approaches are usually more accurate, they can distract the user from the actual analysis task, which may discourage users from recording analytic provenance. The automatic approaches [82] do not introduce interruption to the sense-making process, but their capability for inferring semantic-rich analytic provenance information is limited (see, [77] and Wheat, this volume).

In terms of representation, node-link diagrams are a popular choice among methods that aim to show an overview of the sense-making process [78, 79, 83, 84]. They usually follow the temporal order or the casual relationship among actions. In such methods, nodes represent a summary of system state and the edges represent actions that transit system from one state to another. While providing an overview of the sense-making structure, in many cases node-link diagrams (see Fig. 3) do not have sufficient detail for understanding the semantics of user action. To provide

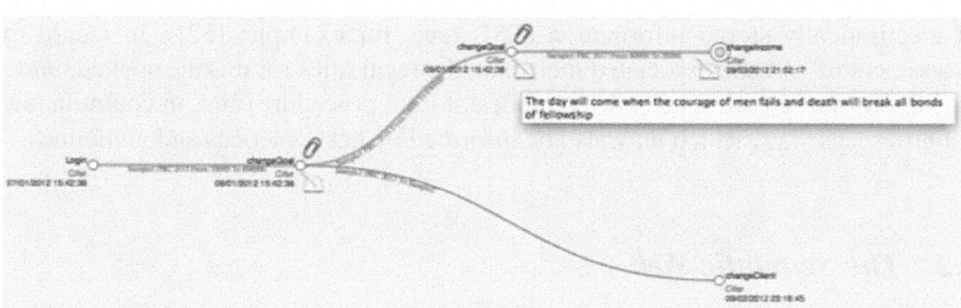

Fig. 3 A node-link representation with note and system state from the "Activity Tracker" of the FIVA—Fixed Income Visual Analytics tool [85]. See also a demonstration video at https://www.youtube.com/watch?v=VGRh0BmJpTY

more context, the most common approach is multiple-coordinated views that show the node and system state only for a selected step [78, 79, 85]. This usually works well with many visual analytics systems, which already have views for each type of information: showing the sense-making context essentially restores the system to a previous state.

6 Provenance Standards and Specifications

As indicated in the previous section, a number of standards and specifications relating to provenance have emerged over the past years, and several of these were discussed at the workshop. Some of these standards and specifications operate at the international level, standardizing practices and technologies for an entire field, while others apply to particular contexts or organizations. Some are concerned with capturing and representing provenance information as a part of information processes and processing, while others are concerned with capturing or assessing provenance information post-hoc as part of forensic or preservation processes, and some comprise a combination of both approaches. In a number of cases these standards consist of data models, data dictionaries, and unique identifiers, with associated ontologies (increasingly represented in OWL format) and registries. This section presents information related to these different types of standards and specifications for each discipline or community of practice, as discussed at the workshop.

6.1 Law

Not all fields have developed standards. In the field of law, for example, there are important best practices and guidelines, such as the Sedona Conference guidelines for managing electronic records and information that include discussion of the capture and preservation of metadata as a means of establishing trustworthiness of electronically stored information (ESI) (see, for example, [52]). In regard to assessment of ESI and associated metadata, the legal rules for this are spelled out in various laws (e.g., criminal codes or codes of civil procedure) and, in common law countries, case law, which may also be informed by best practices and guidelines.

6.2 The Semantic Web

In the context of linked open data on the web, standardization efforts have been led by the W3C effort. These efforts led to creation of the PROV data model and language (see, Missier, this volume, [86], and [87]). The PROV-DM document [33] provides a strong, formal as well as operational definition of provenance for the web community to use and build upon (i.e., provenance as a record that describes the people, institutions, entities, and activities involved in producing, influencing, or delivering a piece of data or a thing [88].

A related concept is Context, which W3C PROV defines as the circumstances that form the setting for an event, statement, or idea, and in terms of which it can be fully understood [33]. A context would include the following: the type of activity; the participants in the activity; the roles of the participants in the activity (if relevant); time of the event, either a time instant or a time duration; location of the event (derived from the location of the participants).

Provenance records are metadata, notes the primer. There are other kinds of metadata that are not provenance. For example, the size of an image is metadata of that image, but it is not provenance information. Activities that are relevant to an object's provenance include: changes in ownership; transfer or declaration of rights; creation of an object; derivation of an object; revision of a digital object; and use of a digital object [88].

The PROV primer goes on to elaborate three different approaches to provenance [88]:

- One perspective might focus on *agent-centered provenance*, that is, what people or organizations were involved in generating or manipulating the information in question. For example, in the provenance of a picture in a news article we might capture the photographer who took it, the person that edited it, and the newspaper that published it.
- A second perspective might focus on *object-centered provenance*, by tracing the origins of portions of a document to other documents. An example is having a web page that was assembled from content from a news article, quotes of interviews with experts, and a chart that plots data from a government agency.
- A third perspective one might take is on *process-centered provenance*, capturing the actions and steps taken to generate the information in question. For example, a chart may have been generated by invoking a service to retrieve data from a database, then extracting certain statistics from the data using some statistics package, and finally processing these results with a graphing tool.

The PROV-O ontology [33, 89], which mirrors the technology-agnostic PROV data model, makes this integration of these different approaches to provenance very natural, by providing the basis for RDF serialization of provenance traces thus enabling, for example, tools such as the generator tool, ProvGen [90] to generate synthetic graphs that respect realistic graphs, for scalability testing of the provenance management infrastructure. Such efforts also make it easier to develop better 'crosswalks' between standards using taxonomy alignment tools. The family of PROV standards is captured in Fig. 4, along with an explanation of each document in Table 1.

There have been many extensions of the PROV standard (e.g., PROV-ONE was developed for the DataONE project), which combined "trace-land" with

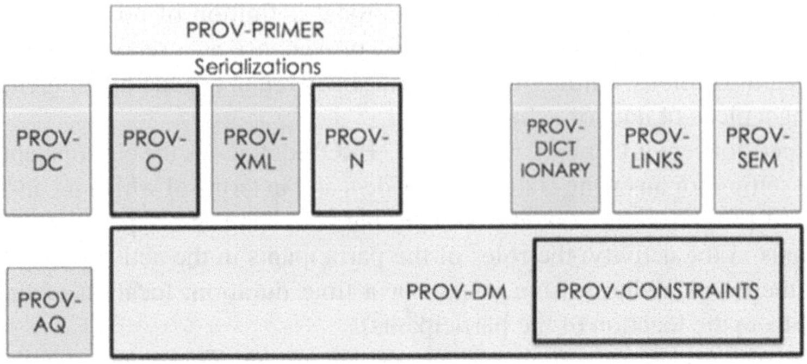

Fig. 4 PROV family of documents [91]

Table 1 Descriptive overview of PROV family of documents [91]

Part	Audience	Type	Document
1	Users	Note	PROV-PRIMER is the entry point to PROV offering an introduction to the provenance data model. This is where you should start and for many may be the only document needed
2	Developers	Rec	PROV-O defines a light-weight OWL2 ontology for the provenance data model. This is intended for the Linked Data and Semantic Web community
3	Developers	Note	PROV-XML defines an XML schema for the provenance data model. This is intended for developers who need a native XML serialization of the PROV data model
4	Advanced	Rec	PROV-DM defines a conceptual data model for provenance including UML diagrams. PROV-O, PROV-XML and PROV-N are serializations of this conceptual model
5	Advanced	Rec	PROV-N defines a human-readable notation for the provenance model. This is used to provide examples within the conceptual model as well as used in the definition of PROV-CONSTRAINTS
6	Advanced	Rec	PROV-CONSTRAINTS defines a set of constraints on the PROV data model that specifies a notion of valid provenance. It is specifically aimed at the implementors of validators
7	Developers	Note	PROV-AQ defines how to use Web-based mechanisms to locate and retrieve provenance information.
8	Developers	Note	PROV-DC defines a mapping between Dublin Core and PROV-O
9	Developers	Note	PROV-DICTIONARY defines constructs for expressing the provenance of dictionary style data structures
10	Advanced	Note	PROV-SEM defines a declarative specification in terms of first-order logic of the PROV data model
11	Advanced	Note	PROV-LINKS defines extensions to PROV to enable linking provenance information across bundles of provenance descriptions

"workflow-land" and a competitor ontology called P-PLAN.[11] One issue with the growing number of extensions is that they are not necessarily logically consistent or interoperable, so the problem of multiple schemas, multiple ontologies, etc. emerges. But despite the inherent issues around standards, having an extension that is agreed upon by a specific community is valuable.

6.3 The Bodleian Library's CAMELOT Data Model: An Implementation of the W3C PROV Standard

The Bodleian Library has developed its own CAMELOT[12] reference Data Model. CAMELOT is a contextual semantic reference data model that frames objects within the scope of the Bodleian Libraries, with context, including provenance. The CAMELOT data model is a description of "real world" objects together with their properties and relationships. The main aim of the data model is to support the development of information systems at the Bodleian Digital Library by providing a human-readable and machine-readable definition and format of data. Implementation of a common data model offers compatibility of data between different systems and the opportunity for data integration in addition to a data model that can be used for reference by those within and outside the organization.

The CAMELOT data model has a modular structure, with each module concerned with the semantics of a particular domain, e.g. types of educational activity. The individual modules are defined using the W3Cs Ontology Web Language (OWL) and consist of entity classes, representing kinds of things of significance in the domain, as well as assertions about relationships between pairs of entity classes. Each of the semantic data models specifies the kinds of facts or assertions that can be expressed using the model, and defines the allowed assertions in a machine-readable language.

The data model captures relationships in context as well as people in context of relationships, as in Figs. 5 and 6.

6.4 Digital Libraries Preservation Metadata Standards

In the field of librarianship, the digital preservation metadata standard PREMIS [93] is increasingly evolving towards a provenance model. The PREMIS Data Dictionary for Preservation Metadata is the international standard for metadata to support the preservation of digital objects and ensure their long-term usability. Developed by an international team of experts, PREMIS is implemented in digital

[11]http://vocab.linkeddata.es/p-plan/.

[12]http://camelot-dev.bodleian.ox.ac.uk/?page_id=20.

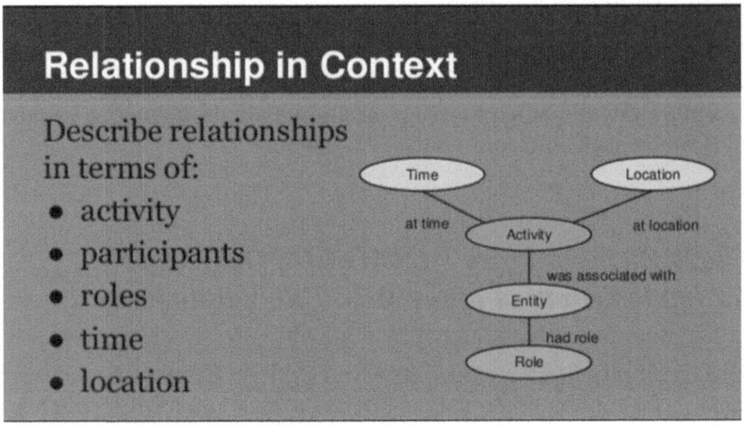

Fig. 5 The Bodleian Library's CAMELOT data model captures relationships in context (Bodleian Library, 2014)

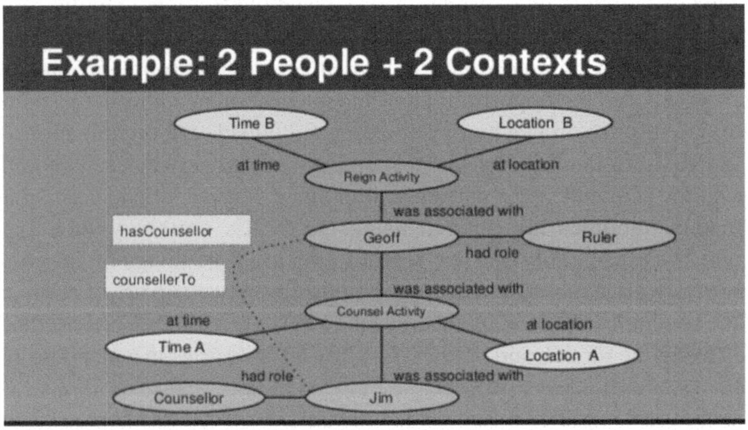

Fig. 6 The Bodleian Library's CAMELOT data model captures people in contexts [92]

preservation projects around the world, and support for PREMIS is incorporated into a number of commercial and open-source digital preservation tools and systems. The PREMIS Editorial Committee coordinates revisions and implementation of the standard, which consists of the Data Dictionary, an XML schema, and supporting documentation. The PREMIS Data Dictionary builds on the Open Archival Information System (OAIS) reference model (ISO 14721) [94]. The OAIS reference model provides a conceptual foundation in the form of a taxonomy of information objects and packages for archived objects, and the structure of their associated metadata. PREMIS can be viewed as an elaboration of the OAIS model, explicated through the mapping of preservation metadata to that conceptual structure. The PREMIS Data Dictionary can be viewed as a translation of the OAIS reference model into a set of implementable semantic units. However, it should be noted that the Data Dictionary and OAIS occasionally differ in terminology usage. Differences usually reflect the fact that PREMIS semantic units require more specificity than the OAIS definitions provide, which is to be expected when moving from a conceptual framework to an implementation. As of 2013, a PREMIS compliant OWL ontology was also available. The OWL ontology, as the PREMIS editorial committee notes, "integrates PREMIS information with other Linked Data compliant datasets, especially format registries, which are now referenced from the PREMIS ontology (for instance, the Unified Digital Format Registry[13]and PRONOM.[14] Thus information can be more easily interconnected, especially between different repository databases. The OWL design of PREMIS should NOT [emphasis in original text] be considered as a replacement for the XML Schema: the two of them should rather be considered complementary. Work to align the PREMIS ontology with the PROV ontology is being considered" [95]. An important caveat about PREMIS is made by Guercio, who notes that, "In the archival environment but also for the dynamic use of the resources as required in the scientific institutions or for performing arts (CASPAR), because of the complexity of records sedimentation and aggregations, the preservation is not only and mainly solved on the basis of a collection of metadata/information (even if very rich like in PREMIS + descriptive metadata)" [96]. Remaining Questions, according to Guercio are how to document and verify the chain of changes before and within the repositories and how to guarantee the maintenance of knowledge accumulated over time by the designated communities [96].

6.5 Digital Records and Archives Preservation Standards

Archives are created when people or organizations perform functions and activities, thus a great deal happens to records before they are transferred to a trusted digital repository in an archives for long-term preservation. This, as Guercio points out above, makes it necessary for archivists and those concerned with the preservation of digital records, as opposed to publications or historical manuscripts, to develop standards for the capture of metadata along the chain of custody, or what some records managers and archivists refer to as the "life cycle" of the records and others (e.g., Australian archivists) refer to as the records "continuum." Cunningham (this volume) discusses an early effort by the Australian archival community to develop a metadata standard for records based on the Australian "continuum" model, represented in Fig. 7.

Work began on an international standard on metadata for records in 2004. The resulting standards, ISO 23081, Information and Documentation—Records Management Processes—Metadata for Records [98]. The standard consists of Part 1

[13]http://udfr.org/onto/onto.rdf.

[14]http://www.nationalarchives.gov.uk/PRONOM/Default.aspx.

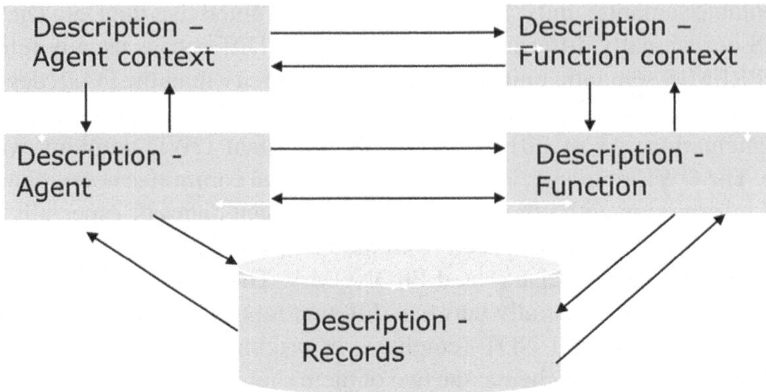

Fig. 7 'Conceptual and relationship models: Records in Business and Socio-legal Contexts', a deliverable from the 1998–1999 Australian Research Council funded Monash University research project, called 'Recordkeeping Metadata Standards for Managing and Accessing Information Resources in Networked Environments over time for Government' [97]

Principles, which sets a framework for creating, managing and using records management metadata and explains the principles that govern them. It is a guide to understanding, implementing, and using metadata within the international standard on records management [100], which addresses the relevance of records management metadata in business processes and the different roles and types of metadata that support business and records management processes. It also sets a framework for managing those metadata. Part 2 [100] Conceptual and Implementation issues, focuses on the framework for defining metadata elements for managing records and provides a generic statement of metadata elements, whether these are physical, analogue, or digital, consistent with Part 1 of the standard. Part 3 [101] comprises a self-assessment framework for organizations to use in assessing their metadata. The data model underpinning the standard is shown in Fig. 8.

The above standards, however, only capture a part of the chain of custody; that is, the part taking place prior to transfer to a trusted digital repository for archival preservation. In contrast, the InterPARES Project has developed a life cycle based model for digital preservation, called the "Chain of Preservation" CoP model [102]. It is defined as "A system of controls that extends over the entire lifecycle of records and ensures their identity and integrity in any action that affects the way the records are represented in storage or presented for use" [103]. As explained by Xie [104], the CoP model is a construct that complements the chain of custody model, extending it into future time. It encompasses all the activities relevant to the preservation of digital records in their authentic form and depicts a complete process. It includes activities typically carried out by both the records creator and the preserver of records. At the highest level, it consists of four major activities: framework management for chain of preservation, records management in a record-making system, records management in a record-keeping system, and records management in a permanent preservation system. Management

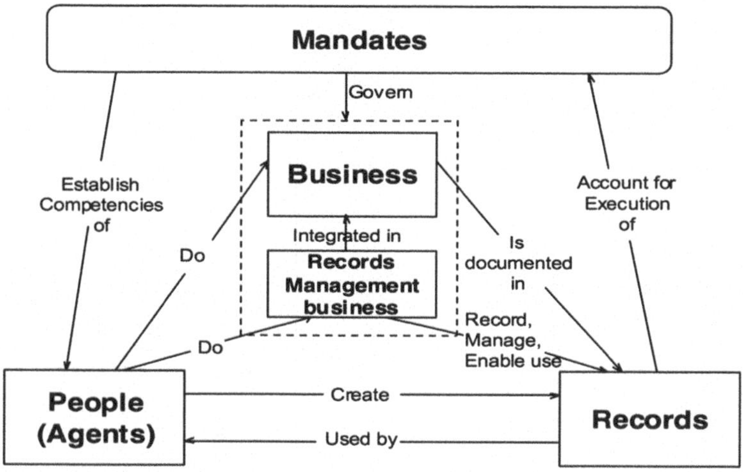

Fig. 8 Data model for metadata for records [99]

for the purposes of the CoP framework includes the design of three systems, which yield products such as policies, procedures, rules, tools, and functional requirements for technological systems, respectively. The management of record making and recordkeeping systems includes the activities performed by records management, and the management of the permanent preservation system includes activities conducted by archival administration. The model, which has been created and exists in IDEF0 format, thus integrates the perspectives of both records creator and the preserver. Figure 9 shows the InterPARES CoP model—Managing Chain of Preservation model.

Subsequent to the period of time records are of use to their creating agencies, those records that are appraised as having enduring value and transferred to agencies concerned with their long-term preservation (i.e., archives), will be "arranged and described". This is a process of enriching previous system and recordkeeping metadata with additional preserver metadata, such as discussed above. A number of standards on archival description have been developed for this purpose.

Without going into a complete survey of these standards and variations in national approaches, the current international family of archival descriptive standards are the International Council on Archives standards for archival description, ISAD (G) in 2000 [16] and ISAAR (CPF) in 2004 [17]. ISAD (G) governs records description, while ISAAR (CPF) governs the description of records creators and their various relationships. The more recent creation by the ICA of a third standard for the description of functions—ISAF [106]—potentially completes the triangle, although arguably more still needs to be done to articulate the complete conceptual model.

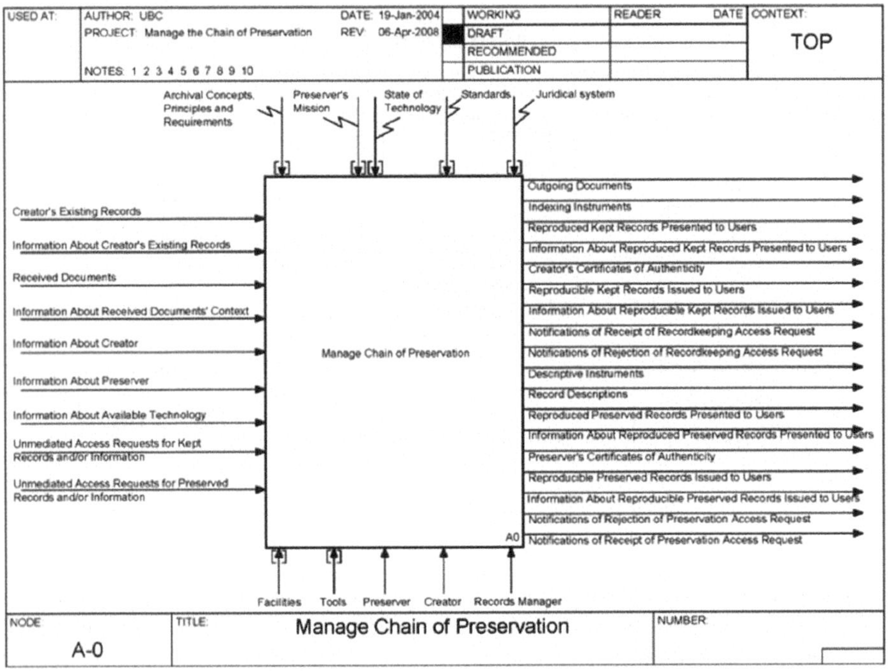

Fig. 9 InterPARES chain of preservation, manage chain of preservation model [105]

6.6 *Visual Analytics*

In visual analytics, the field is simply too young to yet have established a standard approach to capturing or representing provenance information; indeed, how to do this in regard to analytic provenance remains an open and active area of research.

7 Research Challenges

There are a growing number of individuals, groups, and research initiatives concerned with conducting research on provenance. The study of provenance presents a rich field for exploration given the range of open research challenges. This section identifies several that deserve further attention, grouped into those that are conceptual and/or theoretical in nature; those that concern development of interoperability between standards and frameworks; and those that are technical in nature.

7.1 Conceptual/Theoretical

There remain many unresolved aspects of provenance as a concept. A deeper and agreed understanding of provenance is needed in order to cope with new forms of documentation and new modes of communicating and processing information. In addition, from an archival perspective, the most difficult issue is to identify who can be considered the creator of an archival object: organizations change, their denomination is modified, and so too, their organizational assets, along with their mandates and competences.

Similarly, it is possible to question whether a body of records ever has one provenance, and this may be especially true in distributed interconnected environments. Archives reflect and document life and activity in the real world. The real world is a complex place. Relationships in the real world are rarely one-to-one; on the contrary they are usually many-to-many. In the real world archives reflect the complex reality of dynamic inter-relationships between different records-creating entities. A common example of this complexity is the incidence of administrative change in governments and in large corporations. In archival terms this can be understood as successive multiple provenance. But multiple provenances can also occur simultaneously, where more than one entity is simultaneously involved in the creation and use of a given body of archives. This phenomenon has always existed, but is becoming even more prevalent and apparent with digital records, where shared systems often create a single body of archives for multiple separate entities.

In addition, it is necessary to ask, provenance of what? Of archives, records, data, decision-making or analytic processes? Depending on the focus or scope of provenance, there are any number of implications for who the creator is, what it needs to be captured about provenance and how to implement it. Burgess (2015) emphasizes the need for a broader definition of provenance that incorporates the physical/digital transition as well, stating that, for a digital surrogate, the provenance of the physical item and the surrogate will begin to diverge at the moment of digitization. A similar effect happens when a copy of a born-digital artifact is archived. Expressing these multiple histories is an ongoing challenge. Rogers (2015) calls for further exploration of the relationship between provenance and authenticity, which she notes is motivating provenance research in a general sense, while the practical means of preservation is motivating provenance research in a technical context. As it concerns the relationship between provenance and trust, Ken Thibodeau (this volume) calls for empirical evaluation, noting that if we are claiming that provenance improves trust, and that a particular way of representing provenance improves trust, we need to evaluate the claims. We need strategies to assess users' trust in relation to quality of provenance information. Finally, as Missier (this volume) observes, provenance information can be expressed at different levels of abstraction, but open questions remain about what level of granularity and amount of information on provenance is needed in different contexts (e.g., on the basis of users and uses).

Finally, the research agenda on analytic provenance is a novel area of investigation that remains relatively wide open for further examination. Current visual representations need to be enhanced to support more accurate and detailed provenance information, to allow better inference. Varga (this volume) notes that applying visualizations in a narrative context can make complex data more comprehensible, memorable, and credible. With Visual Analytics, these narratives result in being explicit representations of the hypothesis. They often include different types of data in their presentation, such as explicit information from the original raw dataset, provenance data that shows the processes and manipulation to which the data has been subjected, and implicit information from users' knowledge and experience. The challenge is to determine and define the best way to display all this ill-defined information. Roberts et al. [48] distinguish between hard data—that is, explicit knowledge, typically quantitative, from a known source and provenance—and soft data—that is, implicit. Both are crucial to understanding provenance in the context of distributed decision-making.

7.2 Interoperability

As discussed in the previous section on provenance standards, there is a range of different data models, dictionaries, and frameworks relating to provenance. How these models and related standards interoperate remains an unresolved question, and one that, if not answered, could slow progress in addressing gaps in provenance information needed for the use cases discussed in this chapter. Do standards represent all perspectives, functions/roles, and use cases? Certainly, with some archival scholars (e.g., [31] and Thibodeau, this volume) calling for use of new technology and models, traditional archival models and standards could be integrated with the newer semantic models and standards (e.g., RDF, PROV, OPM). Co-operation with different communities is key because even in a single context of use (e.g., digital preservation) the field is populated by a variety of actors and users, all engaging with documentation in same way.

7.3 Technical Challenges

One of the open problems concerns capturing provenance from processes that are distributed over multiple, heterogeneous, autonomous systems. Each of these systems may be expected to provide some fragment of provenance, requiring post hoc composition of these fragments. There is general agreement on the need to articulate systematic methods and technical solutions for capturing and rightly attributing this information.

Another important challenge in today's environment is security of provenance information, which includes assurance that provenance information will remain with the records over time and through technological change. A requirement for accountability, however, is that the provenance itself can be trusted not to have been tampered with. Using provenance traces in, say, a court of law, requires strong non-reputability and security guarantees, which can only be provided by a trusted computing infrastructure [107], where however the notion of trusted provenance does not seem to have been investigated.

With vast networks of interconnected information communication and processing systems, storage and retrieval are bound to be issues that also deserve research attention. In a ‘big data’ world, there is a lot of provenance data. How should this provenance information be stored, indexed, and retrieved? How should it be preserved? What proportions of a potential enormous quantity of provenance information should be preserved?

Wherever there is big data, there is an opportunity for analytics and exploitation of the data. Thus, there is an opportunity to explore how provenance information can be exploited once it has been captured? One novel approach is to use provenance traces about usage of OpenData, to ascribe credit to data contributors. Such credit should take into account multiple generations of data derivation and reuse, and thus requires new models of partial credit to be developed. Missier and his team (see the chapter by Missier this volume) are addressing more technical problems, having to do with efficient storage and querying of provenance graphs on a large scale, and have recently built a provenance generator, ProvGen [90] to create provenance graphs with several million nodes, which can be used to test the performance of provenance management infrastructure. Archivists and digital librarians have begun to exploit the analysis of provenance information for preservation risk assessment and planning, but there is likely many more opportunities to use such data to enhance archival administration and preservation work. Open data formats for provenance information, yet to be developed, would likely facilitate its analysis and exploration.

There is also a need for development of solutions to more easily extract provenance information: Provenance information is readily available in a large number of cases but almost always not in a form immediately amenable to representation (e.g., in PROV-O). Where structured data exists then a mapping can be constructed but with a diversity of formats and the evolution of standards these do require a significant maintenance overhead. Furthermore, much historical provenance information is in prose form or hand/type-written manuscripts that require digitization and/or keying. In the short-term these can be accommodated as annotations in order to establish trust and identity but in prose form can provide limited utility in terms of discovery and analytics.

One of the open problems of “human-in-the loop” cognitive systems concerns capturing provenance from processes that are distributed over multiple, heterogeneous, autonomous systems (machine and human). This is one of the biggest challenges in analytic provenance capture. There is a limited time window to capture such information; even the users themselves may forget what they were doing after a while, at which point it becomes very difficult to recover the analytic provenance information [108]. Users’ knowledge and experience have a considerable impact on the way they conduct analysis. As a result, the sense-making process (i.e. the analytic provenance) can vary significantly from user to user, even with the same dataset and analysis task [108]. An experiment that studied how much of a user's reasoning process can be recovered from user action information [109] showed that the accuracy is not high even in a constrained setting with domain experts doing the inference. Given the diversity of data and analysis involved in the sense-making and the difficulty of replicating expert knowledge/thinking in a computer system, the chance of having a generic technique that can accurately infer semantic-rich analytic provenance information for a variety of analysis tasks is not high [108]. Xu [43] conjectures a promising direction is the development of hybrid or semi-automated approaches for capturing analytic provenance, i.e., mixing the manual and automatic capture to combine their strength. For example, an algorithm that predicts sub-tasks can ask for user feedback (i.e., whether the prediction is correct or not) and use the information to improve itself. Similar approaches can be used to uncover user intention or analysis strategies.

8 Conclusion

The above discussion in this chapter has highlighted the rich complexity and evolution of conceptualizations of provenance, the variety of purposes that provenance services in a wide range of use cases, and the diversity of approaches to the capturing and representing of provenance information used in different fields. This exploration of provenance through an inter- and multidisciplinary lens has shown that the focus of provenance also varies according to domain: In some the focus is on data; in others, records or aggregates of these; in others metadata, and in still others analysis and reasoning. Similarly, some fields refer to provenance as representing context, broadly defined. In others fields, provenance refers to agents of origination while in others, processes and lineage are emphasized. Finally, in some cases a combination of these elements is encompassed in provenance and its representation. This suggests that there is rich territory for exploring more integrated and expansive conceptualizations and definitions of provenance that integrate inter- and multidisciplinary perspectives. There also remain many open research challenges of a theoretical/conceptual and technical nature, or needed to address implementation challenges, such as interoperability of different standards and models to enable provenance-based search, retrieval, and analytics.

References

1. Papritz, J.: Archivwissenschaft. 4 vols. Archivschule Marburg. Institut fur Archivwissenschaft, Marburg (1976)
2. Moreau, L.: The foundations for provenance on the web. Found. Trends Web Sci. **2**(2–3), 99–241 (2010)

3. Yeo, G.: Trust and context in cyberspace. Arch. Rec. **34**(2), 214–234 (2012)
4. Duranti, L.: The odyssey of records managers. In: Burke, F.G., Nesmith, T. (eds.) Canadian Archival Studies and the Rediscovery of Provenance, pp. 29–60. Scarecrow Press, Metuchen (1993)
5. Jones, T.G., Burgess, L., Jefferies, N., Ranganathan, A., Rumsey, S.: Contextual and provenance metadata in the Oxford University Research Archive (ORA). In: Metadata and Semantics Research, pp. 274–285. Springer International Publishing, Berlin (2015)
6. Cohen, F: Digital forensics and electronic discovery. http://all.net (c. 2013)
7. Socha, G., Gelbmann, T.: Electronic discovery reference model. http://www.edrm.net/resources/edrm-stages-explained (2016)
8. Tennis, J.T.: A Kaleidoscope perspective: change in the semantics and structure of facets and isolates in Analytico-Synthetic classification. SRELS J. Inf. Manage. **50**(6), 789–794 (2013)
9. Bearman, D.A., Lytle, R.H.: The power of the principle of provenance. Archivaria. **1**(21), (1985)
10. Schon, D.A., DeSanctis, V.: The reflective practitioner: how professionals think in action. J. Contin. High. Educ. **34**, (1986)
11. Varga, M., Varga, C.: Visual analytics – data, analytical and reasoning provenance. Springer Nature (2016). This volume
12. Duranti, L.: The concept of appraisal and archival theory. Am. Arch. **57**, 328–344 (1994)
13. Lemieux, V.L.: Applying Mintzberg's theories on organizational configuration to archival appraisal. Archivaria. **1**(46), (1998)
14. Cook, T.: Archival science and postmodernism: new formulations for old concepts. Arch. Sci. **1**(1), 3–24 (2001)
15. Pearce-Moses, R., Baty, L.A.: A Glossary of Archival and Records Terminology. Society of American Archivists, Chicago (2005)
16. International Council on Archives. International Standard Archival Description (General). ICA, Paris (1994)
17. International Council on Archives. ISAAR (CPF) International Standard Archival Authority Record for Corporate Bodies, Persons and Families, 2nd edn. ICA, Paris (2004)
18. International Council on Archives. Committee on Best Practices and Standards: Progress Report for Revising and Harmonising ICA Descriptive Standards. ICA, Paris (2012)
19. Duchein, M.: Theoretical principles and practical problems of respect des fonds in archival science. Archivaria **1**(16), 64–82 (1983)
20. Horsman, P.: The last dance of the phoenix or the de-discovery of the archival fonds. Archivaria **1**(54), 1–23 (2002)
21. Gilliland-Swetland, A.J.: Enduring Paradigm, New Opportunities: The Value of the Archival Perspective in the Digital Environment. Council on Library and Information Resources, Washington (2000)
22. Cook, T.: Mind over matter: towards a new theory of archival appraisal. In: Craig, B. (ed.) The Archival Imagination: Essays in Honour of Hugh Taylor, pp. 38–70. Association of Canadian Archivists, Ottawa (1992)
23. Abukhanfusa, K., Sydbeck, J. (eds.): The Principle of Provenance. Report from the First Stockholm Conference on Archival Theory and the Principle of Provenance. Swedish National Archives, Stockholm (1994)
24. Douglas, J.: Origins: evolving ideas about the principle of provenance. In: Eastwood, T., MacNeil, H. (eds.) Currents of Archival Thinking, pp. 23–43. Libraries Unlimited, Santa Barbara (2010)
25. Scott, P.: The record group concept: a case for abandonment. Am. Arch. **29**, 493–504 (1966)
26. Canadian Committee on Archival Description. Rules for Archival Description. Bureau of Canadian Archivists, Ottawa (1990)
27. Barr, D.: The fonds concept in the working group on archival descriptive standards report. Archivaria. **1**(25), 163–169 (Winter 1987–88)
28. Millar, L.: The death of the fonds and the resurrection of provenance: archival context in space and time. Archivaria **1**(53), 1–15 (2002)

29. Nesmith, T.: The concept of societal provenance and records of nineteenth-century Aboriginal-European Relations in Western Canada: implications for archival theory and practice. Arch. Sci. **6**(3–4), 351–360 (2006)
30. Nesmith, T.: Reopening archives: bringing new contextualities into archival theory and practice. Archivaria **60**(60), 259–274 (2006)
31. Lemieux, V.L.: Toward a 'Third Order' archival interface: research notes on some theoretical and practical implications of visual explorations in the Canadian context of financial electronic records. Archivaria **1**(78), 53–93 (2014)
32. EDM Council. Financial Industry Business Ontology. http://www.edmcouncil.org/financialbusiness (2012–2016)
33. World Wide Web Consortium. PROV-O: The PROV Ontology. https://www.w3.org/TR/prov-dictionary/ (2013)
34. DCMI. DCMI Specifications. http://dublincore.org/specifications/ (1995–2016)
35. Buneman, P., Khanna, S., Wang-Chiew, T.: Why and where: a characterization of data provenance. In: Database Theory—ICDT, pp. 316–330. Springer, Berlin, Heidelberg (2001)
36. Cheney, J., Chiticariu, L., Tan, W.-C.: Provenance in Databases: Why, How, and Where. Now Publishers Inc., Breda (2009)
37. Green, T.J., Karvounarakis, G., Ives, Z.G., Tannen, V.: Update exchange with mappings and provenance. In: Proceedings of the 33rd International Conference on Very Large Data Bases, pp. 675–686. VLDB Endowment, Almaden (2007)
38. Davidson, S.B., Boulakia, S.C., Eyal, A., Ludäscher, B., McPhillips, T.M., Bowers, S., Freire, J.: Provenance in scientific workflow systems. IEEE Data Eng. Bull. **30**(4), 44–50 (2007)
39. Amsterdamer, Y., Davidson, S.B., Deutch, D., Milo, T., Stoyanovich, J., Tannen, V.: Putting lipstick on pig: enabling database-style workflow provenance. Proc. VLDB Endowment **5**(4), 346–357 (2011)
40. Thomas, J.J., Cook, K.A.: Illuminating the Path: The Research and Development Agenda for Visual Analytics. National Visualization and Analytics Centre, Richland, WA (2005)
41. Jankun-Kelly, T.J.: The Case for Visual Analysis Provenance Cases, Workshop on Analytic Provenance: Process + Interaction + Insight. CHI (2011)
42. Keim, D.A., Kohlhammer, J., Ellis, G., Mansmann, F.: Mastering the Information Age-Solving Problems with Visual Analytics. Florian Mansmann (2010)
43. Xu, K.: InterPARES Trust Interdisciplinary Workshop on Provenance Participant's Statement. Unpublished document (May, 2015)
44. Pirolli, P., Card, S.K.: The sensemaking process and leverage points for analyst technology as identified through cognitive task analysis. Proc. Int. Conf. Intell. Anal. **5**, 2–4 (2005)
45. Klein, G., Moon, B., Hoffman, R.R.: Making sense of sensemaking 1: alternative perspectives. IEEE Intell. Syst. **4**, 70–73 (2006)
46. Hutchins, E.: Cognition in the Wild. MIT Press, Cambridge (1995)
47. Hollan, J., Hutchins, E., Kirsh, D.: Distributed cognition: toward a new foundation for human-computer interaction research. ACM Trans. Comput. Hum. Interact. (TOCHI) **7**(2), 174–196 (2000)
48. Roberts, J.C., Keim, D., Hanratty, T., Rowlingson, R.R., Walker, R., Hall, M., Jacobson, Z., Lavigne, V., Rooney, C., Varga, M.: From Ill-defined problems to informed decisions. In: EuroVis Workshop on Visual Analytics. Eurographics Association, Geneva (2014)
49. Factor, M., Henis, E., Naor, D., Rabinovici-Cohen, S., Reshef, P., Ronen, S., Michetti, G., Guercio, M.: Authenticity and provenance in long term digital preservation: modeling and implementation in preservation aware storage. In: Workshop on the Theory and Practice of Provenance. ACM SIGMOD **38**(2), 57–60 (2009)
50. Gillean, D., Leveillé, V., Rogers, C.: Records in the Cloud–A metadata framework for cloud service providers. In: Proceedings of the International Conference on Cloud Security Management: ICCSM, p. 166. Academic Conferences Limited, Curtis Farm (2013)

51. Lemieux, V.L., Rogers, C., Thibodeau, K.: InterPARES Trust (international multidisciplinary research into issues of trust in digital objects in online environments) Metadata: Authenticity and Provenance in the Cloud. NATO Specialist Meeting IST-13: Distributed Data Analytics for Combating Weapons of Mass Destruction, Lorton, VA, 15–17 October 2014
52. Sedona Conference: Best Practices Recommendations & Principles for Addressing Electronic Document Production. The Sedona Conference, Sedona (2007)
53. Missier, P., Ludäscher, B., Dey, S., Wang, M., McPhillips, T., Bowers, S., et al.: Golden trail: retrieving the data history that matters from a comprehensive provenance repository. Int. J. Digit. Curation. **7**(1) (2012)
54. Open Data Charter.net: Open Data Charter. http://opendatacharter.net/who-we-are/ (c. 2015)
55. McKinsey Global Institute. Open Data: Unlocking Innovation and Performance with Liquid Information. McKinsey & Co., London (2013)
56. European Union. Data Portal. http://www.europeandataportal.eu (2016)
57. Open Government Partnership. About. http://www.opengovpartnership.org/about (2016)
58. Ballard, M.: Poor data quality hindering government open data programme. Computer Weekly (28 August 2014)
59. Dasu, T., Johnson, T.: Exploratory Data Mining and Data Cleaning, vol. 479. Wiley, New York (2003)
60. Anderson, S.R., Allen, R.B.: Envisioning the archival commons. Am. Arch. **72**(2), 383–400 (2009)
61. Oomen, J., Aroyo, L.: Crowdsourcing in the cultural heritage domain: opportunities and challenges. In: Proceedings of the 5th International Conference on Communities and Technologies, pp. 138–149. ACM, New York (2011)
62. Eveleigh, A.: Crowding out the archivist? Locating crowdsourcing within the broader landscape of participatory archives. In: Ridge, M., Mia Ridge (ed.) Crowdsourcing our Cultural Heritage, pp. 211–212. Ashgate Publishing, Farnham (2014)
63. Dewey, M.: Decimal Classification and Relative Index for Libraries, Clippings, Notes, etc, 8th edn. Forest Press, Tionesta (1913)
64. Trickett, S.B., Trafton, J.G., Saner, L., Schunn, C.D.: I don't know what's going on there: the use of spatial transformations to deal with and resolve uncertainty in complex visualizations. In: Lovett, M.C., Shah, P. (eds.) Thinking with Data, pp. 65–86. Lawrence Erlbaum Associates, Mahwah (2007)
65. Flood, M.D., Lemieux, V.L., Varga, M., Wong, B.L.W.: The application of visual analytics to financial stability monitoring. J. Financ. Stability (2016)
66. Watts, K.A.: Proposing a place for politics in arbitrary and capricious review. Yale Law J. **119**, 2–85 (2009)
67. Kelly, J.E.: Welcome to the era of cognitive systems. http://asmarterplanet.com/blog/2012/05/welcome-to-theera-of-cognitive-systems.html (May 10, 2012)
68. Computing Research Association. Grand Research Challenges in Information Systems. CRA, Washington (2002)
69. Cavelier, K.: InterPARES Trust Interdisciplinary Workshop on Provenance Participant's Statement. Unpublished document (May, 2015)
70. MacNeil, H.: Trusting description: authenticity, accountability, and archival description standards. J. Arch. Organ. **7**(3), 89–107 (2009)
71. Bearman, D.: Description standards: a framework for action. Am. Arch. **52**(4), 514–519 (1989)
72. GBIF (Global Biodiversity Information Facility. What is GBIF. http://www.gbif.org/what-is-gbif (2016)
73. Missier, P., Dey, S., Belhajjame, K., Cuevas-Vicenttín, V., Ludäscher, B.: D-PROV: extending the PROV provenance model with workflow structure. In: Proceedings of the 5th USENIX Workshop on the Theory and Practice of Provenance (TaPP 13). USENIX Association, Berkeley (2013)
74. Murta, L., Braganholo, V., Chirigati, F., Koop, D., Freire, J.: Noworkflow: capturing and analyzing provenance of scripts. In: Provenance and Annotation of Data and Processes, pp. 71–83. Springer International Publishing, Berlin (2014)

75. Lerner, B., Boose, E.: RDataTracker: collecting provenance in an interactive scripting environment. In: Proceedings of the 6th USENIX Workshop on the Theory and Practice of Provenance (TaPP 2014). USENIX, Berkeley (2014)
76. Hertzum, M., Hansen, K.D., Andersen, H.H.K.: Scrutinising usability evaluation: does thinking aloud affect behavior and mental workload? Behav. Inf. Technol. **28**(2), 165–181 (2009)
77. Gotz, D., Zhou, M.X.: Characterizing users' visual analytic activity for insight provenance. Inf. Vis. **8**(1), 42–55 (2009)
78. Shrinivasan, Y.B., van Wijk, J.J.: Supporting exploration awareness in information visualization. IEEE Comput. Graph. Appl. **29**(5), 24–33 (2009)
79. Pike, W.A., May, R., Baddeley, B., Riensche, R., Bruce, J., Younkin, K.: Scalable visual reasoning: supporting collaboration through distributed analysis. In: International Symposium on Collaborative Technologies and Systems, pp. 24–32. IEEE Press, New York (2007)
80. Walker, R., Slingsby, A., Dykes, J., Xu, K., Wood, J., Nguyen, P.H., Stephens, D., Wong, B.L., Zheng, Y.: An extensible framework for provenance in human terrain visual analytics. IEEE Trans. Vis. Comput. Graph. **19**(12), 2139–2148 (2013)
81. Nguyen, P.H., Xu, K., Walker, R., Wong, B.L.W.: SchemaLine: timeline visualization for sensemaking. In: Proceedings of the 18th International Conference on Information Visualization (IV), pp. 225–233. IEEE Press, New York (2014)
82. Gotz, D., Wen, Z.: Behavior-driven visualization recommendation. In: Proceedings of the 14th International Conference on Intelligent User Interfaces, pp. 315–324. ACM, New York (2009)
83. Bavoil, L., Callahan, S.P., Crossno, P.J., Freire, J., Scheidegger, C.E., Silva, C.T., Vo, H.T: Vistrails: enabling interactive multiple-view visualizations. In: Proceedings of IEEE Information Visualization 05, pp. 135–142. IEEE Press, New York (2005)
84. Dunne, C., Henry Riche, N., Lee, B., Metoyer, R., Robertson, G.: GraphTrail: analyzing large multivariate, heterogeneous networks while supporting exploration history. In: Proceedings of the SIGCHI Conference on Human Factors in Computing Systems, pp. 1663–1672. ACM, New York (2012)
85. Lemieux, V.L., Dang, T.: Building accountability for decision-making into cognitive systems. In: Advances in Information Systems and Technologies, pp. 575–586. Springer, Berlin, Heidelberg (2013)
86. Missier, P., Belhajjame, K., Cheney, J.: The W3C PROV family of specifications for modelling provenance metadata. In: Proceedings of the 16th International Conference on Extending Database Technology, pp. 773–776. ACM, New York (2013)
87. Moreau, L., Hartig, O., Simmhan, Y., Myers, J., Lebo, T., Belhajjame, K., Miles, S., Soiland-Reyes, S.: PROV-AQ: provenance access and query. http://www.w3.org/TR/prov-aq (2012)
88. Gil, Y., Miles, S., Belhajjame, K., Deus, H., Garijo, D., Klyne, G., Missier, P., Soiland-Reyes, S., Zednik, S.: PROV model primer. https://www.w3.org/TR/prov-primer/ (2012)
89. Lebo, T., Sahoo, S., McGuinness, D., Belhajjame, K., Cheney, J., Corsar, D., Garijo, D., Soiland-Reyes, S., Zednik, S., Zhao, J.: ProvO: The prov ontology. W3C Recommendation. (2013)
90. Firth, H., Missier, P.: ProvGen: generating synthetic PROV graphs with predictable structure. In: Provenance and Annotation of Data and Processes, pp. 16–27. Springer International Publishing, Berlin (2014)
91. Groth, P., Moreau, L: PROV-Overview. An overview of the PROV Family of Documents. https://www.w3.org/TR/2013/NOTE-prov-overview-20130430/ (2013)
92. The Bodleian Library. CAMELOT: A contextual data model for the Bodleian digital library. http://camelot-dev.bodleian.ox.ac.uk (2016)
93. PREMIS Editorial Committee. Data dictionary for preservation metadata, Version 3.0. OCLC, Washington (2015)
94. ISO/IEC: ISO 14721: 2012– Space data and information transfer systems – Open archival information system (OAIS) – Reference model. ISO, Geneva (2012)

95. PREMIS Editorial Committee. PREMIS OWL Ontology 2.2 now available. https://www.loc.gov/standards/premis/ontology-announcement.html (2013)
96. Guercio, M.: PREMIS and the long-term preservation of complex digital archives: Lessons learned and critical issues from the CASPAR Research. Round Table on PREMIS – Preservation Metadata: Implementation Strategies, Rome Italy (2009)
97. McKemmish, S., Acland, G., Ward, N., Reed, B.: Describing records in context in the continuum: the Australian Recordkeeping Metadata Schema. Archivaria **1**(48), 3–37 (1999)
98. ISO/IEC: ISO 23081: 2006. Information and Documentation – Records Management Processes – Metadata for Records – Part I: Principles. ISO, Geneva (2006)
99. ISO/IEC: ISO 15489: 2001. Information and Documentation – Records Management – Part I: General. ISO, Geneva (2001)
100. ISO/IEC: ISO 23081: 2009. Information and Documentation – Records Management Processes – Metadata for Records – Part 2: Conceptual and Implementation Issues. ISO, Geneva (2009)
101. ISO/IEC: ISO 23081. Information and Documentation – Records Management Processes – Metadata for Records – Part 3: Self-Assessment Method. ISO, Geneva (2011)
102. Duranti, L.: The long-term preservation of accurate and authentic digital data: the INTERPARES project. Data Sci. J. **4**, 106–118 (2005)
103. InterPARES 2 Terminology Database. http://www.interpares.org/ip2/ip2_terminology_db.cfm (2016)
104. Xie, S.L.: Preserving digital records: InterPARES findings and developments. In: Lemieux, V.L. (ed.) Financial Analysis and Risk Management, pp. 187–206. Springer, Berlin, Heidelberg (2013)
105. InterPARES. Chain of preservation model. http://www.interpares.org/ip2/ip2_models.cfm# (2007)
106. International Council on Archives. ISAF: International Standard for Activities-Functions of Corporate Bodies. ICA, Paris (2006)
107. Mitchell, C. (ed.): Trusted Computing. Institution of Electrical Engineers, New York (2005)
108. Xu, K., Attfield, S., Jankun-Kelly, T.J., Wheat, A., Nguyen, P.H., Selvaraj, N.: Analytic provenance for sensemaking: a research agenda. Comput. Graph. Appl. **35**(3), 56–64. IEEE, New York (2015)
109. Dou, W., Jeong, D.H., Stukes, F., Ribarsky, W., Lipford, H.R., Chang, R.: Recovering reasoning processes from user interactions. IEEE Comput. Graph. Appl. **3**, 52–61 (2009)

Part II
Archival Perspectives

Describing Archives in Context: Peter J Scott and the Australian 'Series' System

Adrian Cunningham

Abstract During the 1960s Peter J Scott and colleagues at the then Commonwealth Archives Office (now National Archives of Australia) devised a new approach to archival intellectual control, which separated descriptive information about the creators of records from information about the records themselves. This paper provides an overview of the major features of Scott's system, placing it in its historical context and exploring its impact on the development of international archival descriptive standards.

Keywords Archival description • Australia • Multiple provenance • Series system

1 Peter Scott: Australia's Best Known, but Least Well-Understood Archivist

Peter Scott is arguably Australia's best known, but least well understood archivist internationally. The aim of this chapter is to provide an overview of the major features and underpinning rationale of Scott's system for describing and controlling records in context, placing it in its historical context and exploring its impact on the subsequent development of archival descriptive standards.

2 The Complex Reality of Provenance

As every archivist knows, the thing that separates archives from other forms of information is that they derive their meaning and value from their provenance. If you do not know the provenance of an archival document, then the document can be no more than a decontextualized source of information—an information object that is largely devoid of wider meaning and evidential value. Knowledge of the provenance

A. Cunningham (✉)
Queensland State Archives, Brisbane, Australia
e-mail: adrian.cunningham@archives.qld.gov.au

V.L. Lemieux (ed.), *Building Trust in Information*, Springer Proceedings in Business and Economics, DOI 10.1007/978-3-319-40226-0_2

of a document enables that document to be used as evidence of activities, for it is essential to know who created or received the document and for what purpose. As the international records management standard states, records are:

> information created, received, and maintained as evidence and information by an organization or person, in pursuance of legal obligations or in the transaction of business [1].

One of the main aims of archival description, therefore, is to document this provenance in archival description and in our systems of intellectual control and access. In other words, our archival descriptive systems have to document archives in context. This contextual view of archives is supported by the International Council on Archives, which defines provenance as:

> The relationships between records and the organisations or individuals that created, accumulated and/or maintained and used those records in the conduct of personal or corporate activity [2].

Archival descriptive tools and systems have to document and communicate the relationships between recordkeeping activity and the archives created by persons and organisations. Moreover, documentation of provenance can itself be a useful point of access to archives in archival control systems.

While all archivists agree that provenance is a defining feature of archives, the reality of provenance is, I believe, poorly understood. Many of our descriptive standards and systems are based on the simplistic assumption that there is axiomatically a simple one-to-one relationship between a given provenance entity and a given body of archives. This view was articulated as long ago as 1898 with the publication of the so-called Dutch Manual of Muller, Feith and Fruin [3]. Muller and his colleagues certainly had good reasons for emphasising the importance of not mixing up archives that have different provenance in archival arrangement projects. They had to convince archivists that it was vital to not obscure the provenance of archives by cavalier mixing and sorting. In retrospect, however, it is clear that the rigid adoption of the Dutch rules for arrangement and description led archivists to believe stubbornly that a given body of archives could only ever have one provenance—a belief that, as we shall see, simply does not reflect reality.

Archives reflect and document life and activity in the real world. The real world is a complex place. Relationships in the real world are rarely one-to-one, on the contrary they are usually many-to-many. In the real world archives reflect the complex reality of dynamic inter-relationships between different records-creating entities. A common example of this complexity is the incidence of administrative change in governments and in large corporations. In archival terms this can be understood as successive multiple provenance. But multiple provenance can also occur simultaneously, where more than one entity is simultaneously involved in the creation and use of a given body of archives. This phenomenon has always existed, but is becoming even more prevalent and apparent with digital records, where shared systems often create a single body of archives for multiple separate entities [4].

Given this complex reality, how then should archivists document provenance? First and foremost we should design and build archival systems that reflect rather than distort the complex reality of recordkeeping activity. In a relational database environment this is not a difficult challenge. All that is required is a system that supports separate but linked descriptions of archives and the different entities that create archives. In such systems the data inputs need to be standardized, but the outputs (or the ways in which the inputs can be rendered for human interface and presentation) can be infinitely varied to suit different user requirements. One of the great advantages of computers for archives is that the inputs for our descriptive control systems no longer need to be identical to the user interfaces (or finding aids) to those systems, nor do they need to be constrained by such limited and clumsy tools as card catalogues, calendars and inventories.

3 Evolution of the Australian 'Series' System

Australia is a young nation with an even younger archival profession. When the Dutch Manual was published in 1898 Australia did not even exist as a nation—we had to wait another 3 years for that particular milestone. We had to wait almost 50 years before a national archivist was appointed, albeit as a rather minor functionary within the Parliamentary Library. Indeed, it was not until the 1960s that the archival profession in Australia reached any sort of critical mass. Moreover, we had to wait until 1975 before our archival professional association, the Australian Society of Archivists, was established.

When the Australian Government's Archives Division was established in the 1940s it had the distinct advantage of working with a clean slate. Although the Australian bureaucracy and many of its recordkeeping practices were based on the centuries-old model of the British civil service, our archival control systems had to be built from nothing. Of course, at first the Archives Division was more pre-occupied with identifying records worthy of preservation, rescuing them and placing them in reasonable storage facilities. But by the mid-1950s the Division began to turn its attention to how best to bring these records under intellectual control.

The then Commonwealth Archivist, Ian Maclean, and his colleagues had familiarized themselves with the writings of Sir Hilary Jenkinson and the model of archival practice developed by the Public Record Office in London. Early attempts at achieving intellectual control consisted of trying to impose the so-called 'record group' approach[1] onto the records of the Australian Government. This thinking was reinforced in 1954 when TR Schellenberg of the US National Archives was brought to Australia to advise on the development of our archival systems.

[1]The Society of American Archivists online glossary defines 'record group' as: a hierarchical division that is sometimes equivalent to provenance, representing all the records of an agency and its subordinate divisions. However, the records of a large agency may be broken into several record groups, treating the records of different divisions as separate collections rather than as a series. http://www2.archivists.org/glossary/terms/r/record-group.

While all governments experience administrative change, Australian politicians have elevated it to a fine art. The Australian bureaucratic landscape is an ever-changing one, with the constant reallocation of functions amongst an extremely unstable array of administrative units, government agencies and portfolio departments. While this trend has become more noticeable over time, complex administrative histories have always been a feature of Australian bureaucratic endeavour. When functions are reallocated the records are usually reallocated with them. For example, between 1916 and 1945 the immigration restriction function in the Australian government (and the records documenting the performance of that function) was transferred between ten different government departments: External Affairs; Home and Territories; Home Affairs; Prime Minister's; Markets and Migration; Prime Minister's; Transport; Interior I; Interior II; and Immigration.

It is this problem of multiple provenance that gave Maclean and his colleagues headaches when trying to apply the record group approach to intellectual control. Instinctively, they knew that complex administrative histories required assiduous archival documentation of the context of records creation. They continued with increasing difficulty to try to do this into the early 1960s when a young linguist by the name of Peter Scott was appointed to the Archives. In 1964 Scott made the radical suggestion of abandoning the record group as the locus of intellectual control and instead adopting the function-based series as the means of controlling records [5, 6].

This focus on the record series led perhaps inevitably to Scott's strategies being referred to as 'the series system'. As Chris Hurley [7] and others have since pointed out, however, it was not so much the focus on the series that was the defining feature of Scott's strategies, as it was his insistence on the need to separately document records description and administrative context. Series to Scott provided the most efficient vehicle for documenting records description. As such, series descriptions became free-floating entities that are connected as required to descriptions of all the agencies of government that have contributed to their existence.

Far from being an attack on the principle of provenance, Scott saw his approach as being a more efficient means of documenting the true, and often complex, nature of provenance and recordkeeping systems than is possible using the record group approach. It is the Australian view that provenance cannot be reduced to a simple one-to-one relationship between records creator and records. The simplistic view of provenance, which is embodied in the records group approach to archival description, to us represents a debasement of the archival principle of *respect des fonds*. To many of us in Australia, the record group is more a case of *disrespect des fonds*! Records can, and more often than not do, have multiple provenance relationships, either simultaneously or successively. It behoves us as archivists to design descriptive systems that reflect the dynamic and complex realities of recordkeeping.

In essence the Australian system consists of two inter-related component parts:

- **Context Control**, which is achieved by the identification and registration of records creating and other ambient entities and the documentation of the administrative and biographical histories of those entities, their functional responsibilities and their relationships with each other and with the recordkeeping systems they maintain(ed); and

- **Records Control**, which is achieved by the identification, registration and documentation of record series and/or the items that make up those series.

In the Australian system the contextual entities that need to be documented and linked to descriptions of records include individuals, families, organizations, project teams, government agencies and portfolios, governments themselves, functions and activities. It is the complex web of dynamic relationships between these various entities that underpin the transactions that cause the creation of records. It is therefore essential to capture documentation of these relationships in order to provide the contextual knowledge necessary to understand the content of the records themselves. In Australian continuum thinking—and in the words of my fellow Australian Barbara Reed—records are not seen as 'passive objects to be described retrospectively', but as agents of action, 'active participants in business processes' [8].

As can be seen, the Australian system constitutes a dynamic approach to the intellectual control of records. Using this system any particular set of records can be viewed simultaneously or successively through multiple contextual prisms, thus mirroring the dynamic and contingent nature of records creation itself. The structural elements of the system provide the conceptual and documentary building blocks from which traditional or non-traditional finding aids can be constructed as and when required.

4 Post-Custodialism and the Records Continuum

There is another centrally important feature of the Australian approach to the intellectual control of records. Unlike traditional post-hoc approaches to archival description that focus on the static description of non-current records, the Australian approach can be and is used to achieve intellectual control over all of the records, both current and non-current, in a recordkeeping domain. Right from the earliest days of his appointment, Ian Maclean was committed to the pursuit of an integrated approach to managing all of the records of the Australian government, not just the small subset of records that have been transferred to archival custody.

Under this philosophy of intellectual control, the custodial arrangements under which records are held are no longer of great significance. Certainly it is important to know where records are held at any one time, but they do not have to be in archival custody for the Archives to have a strategic responsibility for, and interest in, bringing them under intellectual control.

In the words of Canada's Terry Cook:

> Scott's approach was to move away from describing records in the custody of an archival institution and arranged there in a single group for a single records creator, and to move

> towards describing the multiple interrelationships between numerous creators and numerous series of records, wherever they may be: in the office(s) of creation, in the office of current control, or in the archives ... Scott's fundamental insight broke through not just the straight-jacket of the record group, but all the 'physicality of archives upon which the record group and so many other approaches to archives are implicitly based. In this way, as is finally being acknowledged, Peter Scott is the founder of the post-custodial revolution in world archival thinking. Although he worked in a paper world, his insights are now especially relevant for archivists facing electronic records, where – just as in Scott's system – the physicality of the record has no importance compared to its multi-relational contexts of creation and contemporary use [9].

And as David Bearman has said, "archivists should find, not make, the information in their descriptive systems" [10]—in other words we should reuse and add contextual value to the metadata dynamically created in the records systems of records creators. This is a very different mindset to that of static post hoc cataloguing, which might be regarded as the traditional approach to archival description.

4.1 What About Functions?

Archives are created when people or organisations perform functions and activities. It is not unreasonable, indeed it is arguably extremely useful, to regard functions as entities in their own right—entities that require separate description with links to both the records that document the function and to the records creators that perform the function [11, 12]. Functions are not mere aspects of the life of a records creating entity—on the contrary records creators such as government agencies can often be regarded as nothing more than episodes in the life of a function [12]. The relationships between the three recordkeeping entities can be illustrated as follows[2] (Fig. 1).

In terms of archival description, this model can be represented as follows (Fig. 2).

[2]Source for Fig. 1: 'Conceptual and Relationship Models: Records in Business and Socio-legal Contexts', a deliverable from the 1998–1999 Australian Research Council funded Monash University research project, called 'Recordkeeping Metadata Standards for Managing and Accessing Information Resources in Networked Environments over time for Government. Commerce, Social and Cultural Purposes', Chief Investigators Sue McKemmish, Ann Pedersen and Steve Stuckey. http://www.sims.monash.edu.au/research/rcrg/research/spirt/deliver/conrelmod.html; model developed by Sue McKemmish, Glenda Acland, Kate Cumming, Barbara Reed, and Nigel Ward. The Australian RKMS was a deliverable from the 1998–1999 Australian Research Council funded Monash University research project, called 'Recordkeeping Metadata Standards for Managing and Accessing Information Resources in Networked Environments over time for Government. Commerce, Social and Cultural Purposes', Chief Investigators Sue McKemmish, Ann Pedersen and Steve Stuckey. McKemmish, S., Acland, G., Ward, N., Reed, B.: Describing Records in Context in the Continuum: The Australian Recordkeeping Metadata Schema. Archivaria 48, 3–43 (1999).

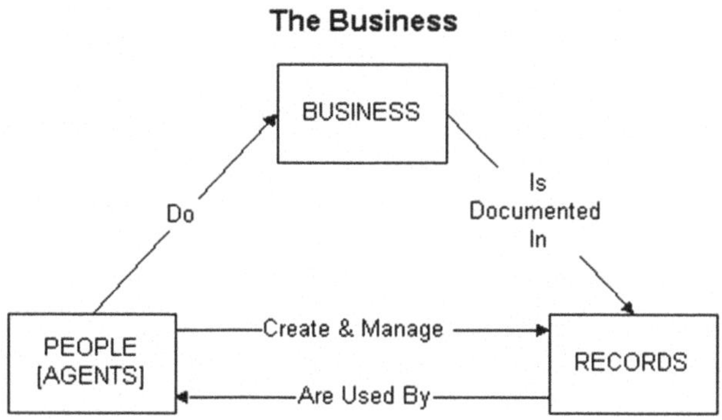

Fig. 1 Relations between recordkeeping entities

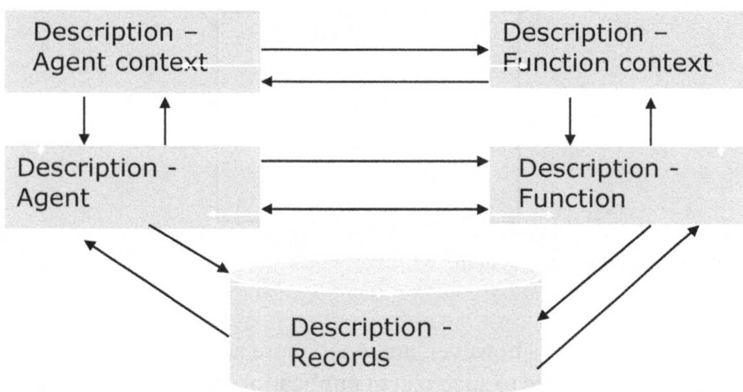

Fig. 2 Series system model of archival description

Within series systems implementations instances of each of the three main entities may be described at different levels of granularity, with relationships between the different levels described accordingly (Fig. 3).

4.2 The Series System and Standards for Archival Description

Those familiar with older guides and standards to archival description would find the Series System to be a very unfamiliar if not incomprehensible approach to intellectual control. I am referring here to such standard sources as the 1898 Dutch Manual, the British *Manual for Archival Description*, the Canadian *Rules for Archival Description* (first edition), the American *Archives, Personal Papers and Manuscripts*, and the first 1994 edition of the *International Standard Archival Description (General)* or ISAD(G) [3, 13–16].

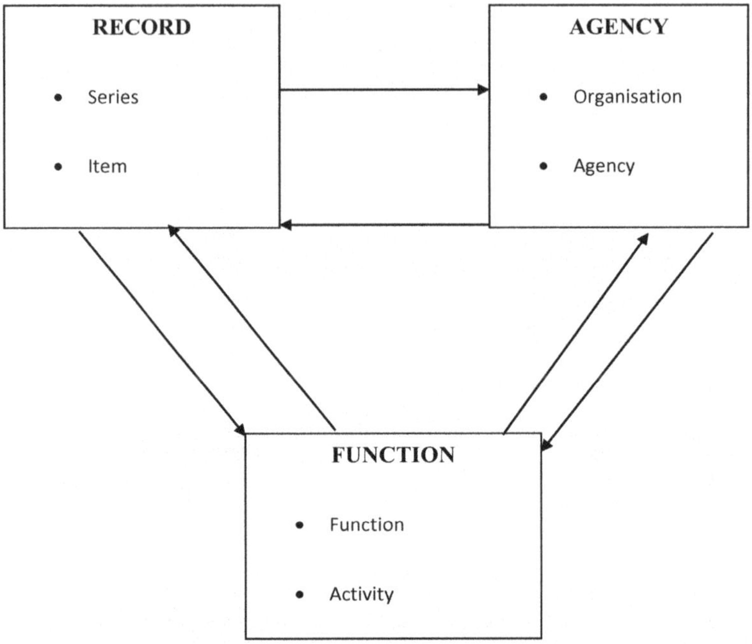

Fig. 3 Relationships in the series system

More recent publications, however, are much more accommodating of the Series System approach. I refer here to such recent publications as the second edition of the *Rules for Archival Description* (RAD2) and the US guide *Describing Archives: A Content Standard* [17, 18]. There has been an international swing towards the logic of having intellectual control systems based on separate but linked descriptions of archives and the context of the creation of archives. Most significant of all has been the publication by the ICA of the second editions of its two companion standards for archival description, ISAD(G) in 2000 and ISAAR (CPF) in 2004 [2]. Very largely, the deployment of these two standards in tandem provides the basis for a series system implementation. Records description is governed by ISAD(G), while the description of records creators and their various relationships is governed by ISAAR (CPF). The more recent creation by the ICA of a third standard for the description of functions—ISDF [19]—potentially completes the triangle, although arguably more still needs to be done to articulate the complete conceptual model.[3]

While Australians have actively contributed Scott's perspectives to the evolution of these international archival descriptive standards, we have not been remiss (though we were perhaps a little slow) in developing our own formal nationally

[3]This work is currently being progressed by the International Council on Archives Expert Group on Archival Description, see http://www.ica.org/en/about-egad

codified archival descriptive standard. This work came to fruition in 2007 when the Australian Society of Archivists' Committee on Descriptive Standards published *Describing Archives in Context: A Guide to Australasian Practice* [20]. Finally practitioners had access to an authoritative and user-friendly reference guide to implementing the series system. This book is still in print and can be purchased from the Australian Society of Archivists.

Archival description has come a long way since Muller, Feith and Fruin and the influence of Peter Scott in that journey continues to reverberate over 40 years since he shared his initial conceptual insights with his colleagues in Australia.

References

1. International Standards Organization: ISO 15489.1 Records Management Part 1: General. ISO, Geneva (2001)
2. International Council on Archives: ISAAR (CPF) International Standard Archival Authority Record for Corporate Bodies, Persons and Families, 2nd edn. ICA, Paris (2004)
3. Muller, S., Feith, J.A., Fruin, R.: Manual for the Arrangement and Description of Archives: Drawn up by the direction of the Netherlands Association of Archivists. Leavitt, A.H.: English translation of the 2nd edn. Society of American Archivists, Chicago (2003)
4. Hurley, C.: Problems with provenance. Arch. Manuscr.: J. Aust. Soc. Arch. **23**(2), 234–259 (1995)
5. Scott, P.: The record group concept: a case for abandonment. Am. Arch. **29**, 493–504 (1966)
6. Wagland, M., Kelly, R.: The series system – a revolution in archival control. In: McKemmish, S., Piggott, M. (eds.) The Records Continuum: Ian Maclean and Australian Archives First Fifty Years, pp. 131–149. Ancora Press, Melbourne (1994)
7. Hurley, C.: The Australian ('Series') System: an exposition. In: McKemmish, S., Piggott, M. (eds.) The Records Continuum: Ian Maclean and Australian Archives First Fifty Years, pp. 150–172. Ancora Press, Melbourne (1994)
8. Reed, B.: Metadata: core record or core business? Arch. Manuscr. **25**(2), 218–241 (1997)
9. Cook, T.: What is past is prologue: a history of archival ideas since 1898, and the future paradigm shift. Archivaria **43**, 17–63 (1997)
10. Bearman, D.: Archival Methods. Archives and Museum Informatics, Pittsburgh (1989)
11. Hurley, C.: What, if anything, is a function? Arch. Manuscr. **21**(2), 208–220 (1993)
12. Hurley, C.: Ambient functions: abandoned children to zoos. Archivaria **40**, 21–39 (1995)
13. Cook, M., Grant, K., Starkey, P.: A Manual of Archival Description. Society of Archivists, London (1985)
14. Canadian Committee on Archival Description: Rules for Archival Description. Bureau of Canadian Archivists, Ottawa (1990)
15. Hensen, S.L.: Archives, Personal Papers and Manuscripts. Library of Congress, Washington, DC (1983)
16. International Council on Archives: International Standard Archival Description (General). ICA, Paris (1994)
17. Canadian Committee on Archival Description: Rules for Archival Description, 2nd edn. Bureau of Canadian Archivists, Ottawa (2008)
18. Society of American Archivists: Describing Archives: A Content Standard. SAA, Chicago (2004)
19. International Council on Archives: International Standard for Describing Functions. ICA, Paris (2011)
20. Australian Society of Archivists: Describing Archives in Context: A Guide to Australasian Practice. ASA, Canberra (2007)

Provenance: An Archival Perspective

Giovanni Michetti

Abstract Archival provenance is a complex concept, the sum of different factors that altogether trace archival records back to their creation and through their management and use. Provenance plays a major role in different archival functions, from arrangement and description to preservation. Therefore, principles and methods for capturing and representing provenance have been developed over a long time in the archival domain. However, further research in this area is needed to cope with the challenges and opportunities of new technology—on the one hand, the digital environment has made it extremely easy to mix and re-use digital objects, to a point that it is often difficult to trace provenance; on the other hand, tools like Resource Description Framework (RDF) can be used to represent provenance through new standards and models.

Keywords Arrangement and description • Digital preservation • InterPARES • Original order • Principle of provenance • Provenance • RDF • Trust

1 Definition and Conceptualization

The International Council on Archives has defined Provenance as

> [t]he relationships between records and the organizations or individuals that created, accumulated and/or maintained and used them in the conduct of personal or corporate activity. Provenance is also the relationship between records and the functions which generated the need of the records [1].

In other words, archival provenance refers to the origins, custody, ownership and use of archival objects. This concept is the basis for the Principle of Provenance—a pillar of Archival Science—which prescribes that archival documents should be arranged according to their provenance in order to preserve their context, hence their meaning.

G. Michetti (✉)
Sapienza University of Rome, Rome, Italy
e-mail: giovanni.michetti@uniroma1.it

V.L. Lemieux (ed.), *Building Trust in Information*, Springer Proceedings in Business and Economics, DOI 10.1007/978-3-319-40226-0_3

The above is a simplification of a complex concept that has been investigated and debated by many scholars since the nineteenth century. In its very early stages, the principle of provenance was mostly meant not to intermingle documents from different origins, that is,

> [r]assembler les différents documents par fonds, c'est-à-dire former collection de tous les titres qui proviennent d'un corps, d'un établissement, d'une famille ou d'un individu, et disposer d'après un certain ordre les différents fonds [2].[1]

However, maintaining the identity of a body of records as a whole is not limited to identifying its distinctness in relation to other records. Archivists soon recognized that the internal structure of such a body also shapes the identity of a fonds, and thus was established the Principle of Original Order—a corollary of the Principle of Provenance. This principle established that groups of records should be maintained in the same order in which they were placed by the records' creator. The underlying idea was that an archives "comes into being as the result of the activities of an administrative body or of an official, and [. . .] it is always the reflection of the functions of that body or of that official" [3].

It was only 50 years ago that such conception was challenged by Peter Scott who—in a seminal article—laid the basis for a further refinement of the principle of provenance: in general, archives are not the result of a single creator who performs a set of specific functions. They are, rather, the outcome of a complex reality where different agents may act as creators; functions change, merge and disappear; and the internal structure is the result of recordkeeping activities that may have little relationship with the business activities of the creators. That is to say, the structure of an archives may have little or no correspondence with the structure of the creating organization. This approach led to a new understanding of the concept of provenance as it is now understood and accepted by the archival community—a network of relationships between objects, agents and functions.

In recent years, the meaning of provenance has been investigated further, and new perspectives have been proposed:

> The similar notions of societal, parallel, and community provenance have also been advanced. They reflect an increasing awareness of the impact of various societal conditions on records creators and record creation processes at any given time and place across the records' history. [. . .] Some archivists have broadened the concept of provenance to include the actions of archivists and users of archives as formative influences on the creation of the records [4].

In particular, Tom Nesmith has provided a definition of provenance that—while giving rise to some issues due its very broad scope—may provide a basis for a broadened multidisciplinary perspective on provenance:

> The provenance of a given record or body of records consists of the societal and technical processes of the records' inscription, transmission, contextualization, and interpretation, which account for its existence, characteristics, and continuing history [5].

---

[1] Transl.: Aggregate all different records in fonds, that is, group all the documents coming from the same body, institution, family or individual, and set the different fonds according to a certain order.

In conclusion, archival provenance is a complex concept, the sum of different factors that altogether trace archival records back to their creation and through their management and use.

2 Relationship to Current Research

This chapter now turns to discussing the author's current research, which has a close relationship with the concept of provenance and focuses on these areas:

- **Trust and digital records**. The author is a member of the InterPARES Trust research project, aimed at generating the theoretical and methodological framework needed to develop policies, procedures and regulations concerning digital records entrusted to the Internet, to ensure public trust grounded on evidence of good governance, and a persistent digital memory. Provenance is a crucial factor of evaluation when assessing the credibility of records on the Internet, therefore provenance needs to be investigated in order to shed light on the nature and the dynamics of the relationship between trust and provenance.
- **Digital preservation**. InterPARES supports a number of research projects, and one of these is PaaST (Preservation as a Service for Trust), which is concerned with investigating digital preservation in the Cloud. The aim of this team is to design a model and a set of functional requirements for preservation of digital records in the Cloud, in order to provide insight and guidance to both those who entrust records to the Internet and those who provide Internet services for records. Preservation, including digital preservation, is about keeping objects along with the context that provides meaning to them. Provenance plays a major role in identifying and determining such context, hence supporting the definition of the identity of the objects targeted for preservation. In addition, provenance of digital objects is itself a digital object that also requires preservation. Both provenance and provenance of provenance are fundamental aspects in any preservation model, theory and practice.
- **Arrangement and description**. Archival arrangement and description entails the creation of representation models in the archival domain. With a growing number of records being created and preserved using Cloud technology, there is a need to consider how to undertake their arrangement and description in the Cloud. Thus, InterPARES is also supporting research aimed at investigating how the Cloud environment may possibly affect arrangement and description theory and practice. Information on provenance is crucial in order to determine the creator of archival materials and identify records' chain of custody, which in turn affect the way materials are arranged and subsequently described. Thus, provenance has an impact on arrangement and description. At the same time, representation models affect the way provenance is understood and represented in archival descriptions, because they highlight certain features while hide or obfuscate others. In short, provenance is a crucial dimension of any arrangement and description process.

- **Linked Data**. Archives are no more made by simple, static documents in the traditional form of a written text on a piece of paper. Organizations and individuals—e.g., researchers—create and publish sets of open data that are then used, mixed and re-used. This raises an issue with regard to the reliability and authenticity of such data, which needs reliable and authentic information on provenance in order to be managed.

3 Motivations for Research

Provenance plays a major role in different archival functions:

- **Preservation** requires maintenance of the context, that is, the complex network of relationships—along with the system of their meanings—in which archival objects have been created, managed and used. Provenance is by definition a crucial part of this context, because even its narrowest definition will address creation and custodial history (i.e., the chain of agents that held the materials, along with related facts and events).
- **Arrangement and description** requires identification and proper description of both the creators and the chain of custody of archival materials. When arranging, provenance is the first clue to trace archival materials back to their origins, identify different bodies of materials, and get to a first, approximate grouping. When describing, the complexity of provenance may affect the representation of the archival materials—this is indeed more true in the digital realm, where new visualization tools and information models allow for greater freedom when designing archival descriptions. Moreover, materials on the Internet are not only dispersed but also mixed and re-used to a point that it is often difficult to trace provenance, hence to trust an archival resource. Some investigation is needed to understand whether traditional concepts and methods can be applied to identify and manage provenance on the Internet, thereby supporting proper arrangement and description of materials.
- **Access and use** of archival materials is both welcomed and actively promoted by archivists. Provenance plays a role when accessing archival materials, since it is one of the key access points—in fact, the names of either the creator or the institution holding the archival materials are among the most common elements used in archival queries. Given a situation in which provenance is more and more a complex network of relationships—if not a confused tangle—it becomes important to allow users to understand such complexity without overwhelming them with a mass of information. Archivists are mediators—as such they have to provide a perspective. Archival representations of provenance in the form of descriptive finding aids form a major part of this perspective—that is why provenance needs to be thoroughly investigated.

- **Appraisal** is the process of assessing the value of records for the purpose of determining the length and conditions of their preservation. According to a widespread approach (known as macro-appraisal), this archival function should be based on "extensive research by archivists into institutional functionality, organizational structures and work-place cultures, recordkeeping systems, information workflows, recording media and recording technologies, and into changes in all these across space and time" [6]. Provenance covers several of these factors, once we assume that it is more than just origination. Therefore investigation on the concept of provenance may have a direct impact on appraisal methods and principles.
- **Technology** is not an archival function, however it is worth mentioning as a motivation for research on provenance, because it affects the way archival functions are interpreted and carried out. In particular, the extended adoption of the RDF[2] model and the general trend towards open government are changing the archival scene and impacting on objects and actors: datasets and distributed computing have entered the archival landscape, while IT specialists have started working on provenance from their perspective, developing their own principles, methods and standards. Therefore, it is important that archivists join the broader discussion bringing the archival voice to the table.

4 Capturing and Representing Provenance

Provenance of archival materials can be captured—most usually manually—from various sources. First of all, a diplomatic analysis[3] of the materials is the fundamental step to identify creators and any other agents that have had some relevant interactions with the materials. Then, reports, accession registers,[4] finding aids[5] and any other document recording information on the creation, management and use of the archival materials may help in reconstructing its custodial history. Direct witness from any agents (creators, managers, archivists, users) may also be of assistance. The biography of the individuals, or the administrative history of the organizations that created and/or managed the materials along with information about their

[2]Resource Description Framework.

[3]Diplomatic analysis is the critical examination of a record carried out on the basis of the principles and methods of Diplomatics. Diplomatics is the discipline that studies the form of written documents (i.e., their logical and physical characteristics) along with their genesis and textual tradition (i.e., how they came into being, and how they have been modified since their creation).

[4]An accession register is an administrative record documenting the process of transferring materials to a repository. It contains key information about the archival materials that have been taken into the physical custody of an archives.

[5]A finding aid is any description providing physical and intellectual control over archival materials, thus assisting users accessing and understanding the materials.

mandates and competences, also aids understanding of provenance. Knowledge of the history of the period during which archival materials have been created, managed and preserved put them in a broader historical context. The physical characteristics of the materials may be of some help as well. In the digital environment, metadata associated with or embedded into materials may provide relevant information on the provenance of either the materials themselves or the systems in which they reside. If the scope of provenance is broadened to include societal provenance,[6] the list of sources needs to be extended to include materials documenting aspects of both the society at large and the specific communities in which the materials have been created, managed and used.

Provenance is usually represented in finding aids in the form of either narratives in textual documents or data elements in software applications. Description should be carried out according to national or international standards, not only for the purpose of interoperability, but also because they usually include specific information elements conveying information on provenance. Even so, such information may be dispersed through different metadata elements or the model may not represent adequately the complexity of concepts like provenance and authenticity, as some scholars have suggested [7]. In recent years, new technology has pushed archival description towards redefinition of the traditional approach. RDF allows for an atomic fragmentation of data elements that can then be aggregated and represented adopting visualization techniques and strategies (e.g., graphs and graph exploration) never used before in the archival domain, dominated by written word, narrative and hierarchical diagrams. This opens up new opportunities for representing the complex network of relationships underlying—rather, making up—an archives, including the possibility of capturing additional layers of provenance in an automatic or semi-automatic way. At the same time, RDF poses new challenges, since it can be used to represent provenance through standards and models (e.g., PROV Ontology [8]) that are not specific to the archival domain, thus requiring a joint effort of different communities to develop shared solutions.

5 Research Challenges

The key challenge in establishing archival provenance is the identification of the creator. Organizations change, their denominations are modified, and so do their organizational assets, along with their mandates and competences. Archivists may have a very clear picture of what happened; nevertheless, they may have difficulties in deciding who the creator is because such decision depends on a discretional evalu-

[6]Societal provenance is a term used to mean provenance in the broader sociocultural dimension. Records creation, management, use and preservation are sociocultural phenomena. Therefore, provenance should be interpreted taking into account the sociocultural dimension as the context in which all actions take place.

ation of the extent and depth of the changes [9]. The same is true for personal papers: there are no organizational assets to worry about, and changes of denomination are not the norm; however, individuals usually organize their records with more freedom than in a corporate environment. As a result, it may be difficult to establish the boundaries between the family archives, the archives of each individual belonging to the family, and the archives of the companies they were possibly holding. This happens because the principle of provenance is, indeed, uncomplicated and agreed in its very basic form (i.e., materials coming from different creators do not have to be mixed), but when it comes to its implementation is not always easy to implement because of the challenges associated with distinguishing whether an entity has died and a new entity has taken its place or it is the same entity that is just growing and re-shaping. As a result, identifying the creator, thus provenance, may be a hard challenge—as Duchein puts it, "[l]ike many principles [. . .] it is easier to state than to define and easier to define than to put into practice" [9].

A more general issue is that there is no consensus within the archival community on the concept of provenance—some still think of it as referring to creation only; others include the custodial history of archival material in its scope, while more recent interpretations have taken into account communities and societies at large [10]. The approach proposed by Peter Horsman may serve to establish a common view. According to Horsman [11], the principle of provenance has an outward application, that is, it functions as a way to identify a body of archival materials as created by a certain creator (individuals, families, organizations), hence separated and distinguished from any archival materials in a repository or elsewhere. The principle has an inward application too, that is, it functions as a method to identify the internal structures of a body of materials, recreating the so-called original order. The key point is to identify the creators and recognize the different roles of any actor who dealt with the materials, i.e., managed, collected or used them. This is a fundamental step, because in the simplest case there will be a creator along with a chain of custody representing the story of different entities holding, managing, using and preserving the materials. In the most difficult cases, despite Duchein's theorization it may be hard to distinguish who can be considered the creator of a complex archival fonds. Therefore, it is important to recognize the role and the contribution of all the entities that dealt with the materials.

In this regard, RDF may be key to the definition of an information model supporting different perspectives on provenance. RDF triples can be used to express specific types of relationships and establish different connections among entities. There would be no need to agree that certain elements are integral to provenance and to reject certain others, the story could simply be told, and the model for telling it could be made sufficiently compassing to allow everyone to tell their stories.

Another research challenge associated with provenance is the clear identification of some mechanisms by which it can support trust in a digital environment. There is no consolidated definition of trust in the archival domain—InterPARES Trust is working to this aim. However, it is agreed that trust is a multifaceted concept based on confidence, vulnerability and risk. Trusting an archival object has to do with the belief that such object can be relied upon. Such reliance is usually the result of a risk assessment—conducted either intentionally or not—where the significant properties of the object itself are analyzed and assessed. Provenance is one of the most meaningful properties contributing to such assessment; therefore, it contributes significantly to the trust-making process. However, besides abstract considerations, no analytic model, methods or metrics have been designed and implemented to support the evaluation of reliability of digital objects on the basis of information on their provenance. Prior to the digital era, archival materials were trusted because of their placement within a trusted repository, i.e., an archives, with preservation, access and use of documentary objects taking place in an environment or according to processes that were considered trustable. The digital environment has corrupted such belief. The challenge is to do something similar to what has been done with markup languages, i.e., making explicit what is implicit. Archivists and records managers need to retain control of provenance and make it explicit, so that users are aware of the quality of the objects and trust them accordingly. The challenge is to find models, mechanisms and tools to achieve this aim, solid enough to meet scientific criteria, but easy enough to be managed by users.

In general, use of new technology and models is another challenge, since it means that traditional archival models need to be compared and possibly integrated with the emerging ones. In this regard, co-operation with diverse communities is key, because the scene is populated by a variety of actors and users, all engaging with the same documentation, but possibly using domain-specific approaches.

In conclusion, the fundamental topic that should be investigated may be: interoperable models to govern and represent provenance in a cross-domain environment. This is an umbrella theme under which different sub-themes may be investigated, such as: granularity and amount of information on provenance based on users' needs and practices; characteristics of existing models of provenance; strategies to assess users' trust in relation to the quality of information on provenance; and analyses of case studies.

Appendix: Bibliography[7]

In order to acquaint those outside the field of archival science with archival thinking on provenance, what follows is a brief bibliography on the topic.

Selected Works

- Cook, T.: What is Past is Prologue: A History of Archival Ideas Since 1898, and the Future Paradigm Shift. Archivaria 43, 17–63 (1997)

[7]Titles already included in the References are not listed in the Bibliography.

- Schellenberg, T.R.: Modern archives: Principles and techniques. University of Chicago Press, Chicago (1956)
- Scott, P.J.: The record group concept: A case for abandonment. American Archivist 29, 493–504 (1966)

Short Bibliography

- Abukhanfusa, K., Sydbeck, J. (eds.) The Principle of Provenance: Report from the First Stockholm Conference on the Archival Principle of Provenance. 2–3 September 1993. Swedish National Archives, Stockholm (1994)
- Boles, F.: Disrespecting Original Order. American Archivist 45, 26–32 (1982)
- Bearman, D.A., Lytle, R.H.: The Power of the Principle of Provenance. Archivaria 21, 14–27 (1985–186)
- Brothman, B.: Orders of Value: Probing the Theoretical of Archival Practice. Archivaria 32, 78–100 (1991)
- Douglas, J.: Origins: Evolving Ideas about the Principle of Provenance. In: Eastewood, T., MacNeil, H. (eds.) Currents of Archival Thinking, pp. 23–43. Libraries Unlimited, Santa Barbara, CA (2010)
- Horsman, P.: The Last Dance of the Phoenix, or The De-discovery of the Archival Fonds. Archivaria 54, 1–23 (2002)
- Hurley, C.: Problems with provenance. Archives and Manuscripts 23, 234–259 (1995)
- Millar, L.: The Death of the Fonds and the Resurrection of Provenance: Archival Context in Space and Time. Archivaria 53, 1–15 (2002)
- Posner, E.: Max Lehmann and the Genesis of the Principle of Provenance. In: Munden, K. (ed.) Archives and the Public Interest: Selected Essays by Ernst Posner, with a new introduction by Angelika Menne-Haritz, pp. 36–44. Society of American Archivist, Chicago (2006)
- Sweeney, S.: The Ambiguous Origins of the Archival Principle of "Provenance". Libraries & the Cultural Record 43, 193–213 (2008)
- Yakel, E.: Archival Representation. Archival Science 3, 1–25 (2003)
- Yeo, G.: Custodial History, Provenance, and the Description of Personal Records. Libraries & the Cultural Record 44, 50–64 (2009)

References

1. International Council on Archives: ISDF. International Standard for Describing Functions. First Edition. Developed by the Committee on Best Practices and Standards. Dresden, Germany, 2–4 May 2007. International Council on Archives, Paris (2007)
2. Instructions pour la mise en ordre et le classement des archives départementales et communales. Paris, 24 avril 1841. In: Lois, Instructions et Règlements Relatifs aux Archives

Départementales, Communales et Hospitalières, pp. 16–28. H. Champion, Libraire, Paris (1884)
3. Muller, S., Feith, J.A., Fruin, R.: Manual for the Arrangement and Description of Archives, 2nd edn., trans. A.H. Leavitt. Society of American Archivists, Chicago (2003)
4. Nesmith, T.: Principle of provenance. In: Duranti, L., Franks, P. (eds.) Encyclopedia of Archival Science, pp. 284–288. Rowman & Littlefield, Lanham (2015)
5. Nesmith, T.: Still fuzzy, but more accurate: some thoughts on the "Ghosts" of archival theory. Archivaria **47**, 136–150 (1999)
6. Cook, T.: Macroappraisal in theory and practice: origins, characteristics, and implementation in Canada, 1950–2000. Arch. Sci. **5**, 101–161 (2005)
7. MacNeil, H.: Trusting description: authenticity, accountability, and archival description standards. J. Arch. Organ. **7**, 89–107 (2009)
8. World Wide Web Consortium: PROV-O: The PROV Ontology. W3C Recommendation 30 April 2013, http://www.w3.org/TR/2013?REC-prov-o-20130430/
9. Duchein, M.: Theoretical principles and practical problems of Respect des Fonds in archival science. Archivaria **16**, 64–82 (1983)
10. Bastian, J.A.: Reading colonial records through an archival lens: the provenance of place, space, and creation. Arch. Sci. **6**, 267–284 (2006)
11. Horsman, P.: Taming the elephant: an orthodox approach to the principle of provenance. In: Abukhanfusa, K., Sydbeck, J. (eds.) The Principle of Provenance: Report from the First Stockholm Conference on the Archival Principle of Provenance, 2–3 September 1993, pp. 51–63. Swedish National Archives, Stockholm (1994)

Research Issues in Archival Provenance

Kenneth Thibodeau

Abstract This chapter reviews the conceptualization of archival provenance and the related concepts of archival fonds and original order and the ways these concepts have guided practice. Section 1 identifies problems entailed by the qualitative and imprecise conceptual approach traditionally applied to the issues involved in the management, use, and preservation of records, and suggests a multidisciplinary strategy to reformulate basic concepts. Section 2 describes problems arising from the traditional approach in greater detail. Section 3 indicates how a multidisciplinary strategy might be applied to both clarify theory and improve practice.

Keywords Archival science • Graph theory • Original order • Provenance • Systemic functional linguistics

1 Introduction

The management of records and their preservation in accordance with the dictates of archival science and well-established practices has been guided by concepts articulated in a purely qualitative, largely philosophical, and often rhetorical mode. In a world where the information used in the conduct of affairs is increasingly encoded in digital bits and where digital information is growing exponentially [1], the conceptual foundations for managing records, both for current business and to enable exploitation of their long term value for diverse and often unforeseeable purposes, need to be reformulated with greater precision in representation, clarity in distinctions, and verifiability in implementation. This reformulation is needed to ensure that sound concepts and methods are implemented unambiguously and effectively, to increase the adaptability and efficacy of the governance of institutional information, and to respond to the challenges posed by the continuing stream of new forms of information and new ways of communicating and using information enabled by information and communication technologies (ICT).

K. Thibodeau (Retired) (✉)
National Archives and Records Administration, College Park, MD, USA
e-mail: KThibodeau@Fordham.edu

V.L. Lemieux (ed.), *Building Trust in Information*, Springer Proceedings in Business and Economics, DOI 10.1007/978-3-319-40226-0_4

This chapter explores the potential for reformulation of the basic concepts that guide the management of records and archives through the application of systemic functional linguistics and mathematical graph theory.

2 A State of Confusion

Currently, the concepts that guide management, preservation and communication of records suffer from a combination of narrowness and vagueness. Perhaps surprisingly, the one follows from the other. Archival thinking and practice are constrained by the two overarching and complementary concepts of provenance and original order. While there are substantial disagreements about what the two concepts entail, there seems to be universal acceptance of their fundamental importance. Reverence towards the two concepts leads those who perceive issues with either their conceptualization or their implementation to focus on broadening the concepts, with the net result that they are overloaded to the point of becoming confusing and impractical.

Problems with the traditional conceptual approach to managing and preserving records begin at the most basic level. They include disagreements about both the intensive and extensive definition of 'record;' confusion between the quality of a record and its existence; failure to distinguish categorically or consistently the properties of the things that are designated as records from the properties of that which makes them records; overly zealous application of definition to justify the exclusion of documents from the application of records management controls that would serve organizational interests; and the mutation of successful techniques for addressing basic problems of managing records into abstract concepts that are proffered as essential requirements.

In principle, any type of persistent information object can be a record; moreover, an information object could be a record in one context and not in another. Similarly, the same object can be different records in different contexts. The criteria that determine whether an information object is a record are independent of the characteristics of the object itself. The properties of a document as such can make it a better or worse record, but they cannot make it a record in the first place. The key criteria that determine whether any given object is a record are (1) a record is a document that is used in an activity of the person or organization whose record it is and (2) it is kept, ideally under records management control. These criteria are often stipulated in laws, regulations, standards and policies related to records [2–4]. For example, the U.S. Federal Records Act defines records as "all recorded information, regardless of physical form or characteristics, made or received by a Federal agency under Federal law or in connection with the transaction of public business and preserved or appropriate for preservation . . . ." [5].

Thus, there is no such thing as a record pure and simple. It must be a record of something: some act, activity, or state of affairs. What makes a document a record is the link between the document and the context in which it was used. This link may not be evident in, or even reliably inferable, from the document itself. For example, one would infer from its content that a written contract for the acquisition of technical services by a government agency is a record of the agency's procurement activity; however, that is not necessarily the case. If the contract were found in the case files of a law enforcement agency, it would be a record of an investigation of some alleged illegal activity, such as bribery of a government official. The different contexts in which a document can be a record may even be independent of one another. Records created for weather forecasting can become important in criminal cases; for example, they could be used to impugn eye-witness testimony if they showed that, in spite of a full moon, there were dense clouds and therefore poor visibility the night a crime occurred.

The fact that a record is defined not by what it is intrinsically, but by its relationship to activity creates an anomalous situation with respect to the provenance of a record because traditionally archival provenance has been limited in scope to the records of a single records creator or a succession of records creators [6, 37]. Even when a broader scope is adopted, the provenance of records does not encompass actual creation of documents that are received from outside sources, such as incoming correspondence and reports.

In archival science and practice, provenance has both internal and external dimensions. Externally, provenance is delimited by the archival fonds, conceived as the totality of records created by a single records creator. Internally, provenance reflects the relationships between records and the activities in which they were instrumental. This reflection is seen as embodied in the way records were organized by their creators. Both dimensions are the cornerstones of archival theory and practice epitomized in the principles of *respect des fonds* and respect for original order [1]. Provenance encompasses the relationships between records and the entities that create, keep or use them [7]. Original order is "The organization and sequence of records established by the creator of the records" [8]. There are several problems with the ways both provenance and original order have been conceptualized and used.

To explain, archival provenance is determined when a document is captured or set aside as a record. This appears reasonable in the context of traditional hard-copy records. When records are inscribed on physical media in a hard and fast manner and placed in a physical folder, there is little likelihood that they would be changed, either in themselves or by relocation to a different folder. In this environment, information about who made the document a record in what activity could be a sufficient description of its provenance. However, even for hard copy records, the concept is overly restrictive. As the Australian concept of records series emphasizes, records can be used in the same activity by more than one actor. Governments, for example, carry out specified functions for as long as the laws that authorize or require them are in effect. But during that time government entities are often reorganized, resulting in different organizations carrying out the same function successively, often using the same records, or the same types of records, arranged in the same way [9–12]. The definition of provenance as a univocal and unalterable attribute of a set of records needs to be modified to reflect these and similar situations.

A second aspect of the arbitrary narrowness of the concept of record provenance is that it is bounded by the recordkeeping of an actor or succession of actors. In most cases the records of any given creator include a large proportion of documents that were not produced by the records creator, but acquired from other sources in a variety of ways. Records received from external parties, such as incoming correspondence, are obviously shaped by their authors. Records created in response to received records are also influenced by the authors of the incoming documents. The extent and limitations of acquired records, whether received from outside parties acting independently or purposely acquired at the initiative of the records creator, can also influence the records creator's decisions on whether it needs to acquire or produce other records to carry out an activity and, if so, what should be the form and content of the additional records.

Even when a records creator has extensive control over the submission of documents, their authors still have a decisive role. For example, the U.S. National Institutes of Health (NIH)[1] dictates the form of applications for research grants, publishes formal guidance on applying for grants, provides an online system for development, submission and tracking of applications, and publishes notices identifying specific areas where program managers believe that there is both a need for new knowledge and a significant probability of successful research [13–15]. Nonetheless, the specific subjects, methods, scope, and other details of the research projects proposed in grant applications are determined by the competence and creativity of the applicants, and shaped by factors such as their understanding of the state of science, their estimates of likely competition and alternatively of possible collaborations, the research resources available to them, the influence of their mentors and peers, the availability of other sources of funding, and even their knowledge of the government and external experts likely to be involved in reviewing their applications. The importance of researcher initiative and insight is reflected in the longstanding belief held by NIH managers that the best research results from ideas formulated by the researchers themselves, rather than by government officials [16]. Thus the substance of the research conducted with NIH grant funding is substantially determined by the applicants and their success or failure influences the subsequent direction of the research funding. Thus, limiting provenance to the records of the NIH arbitrarily truncates any effort to understand how and why the records are what they are. This observation can be extended to most contexts in which records are created, except in the uncommon and mostly uninteresting case of activities carried out entirely by one party.

An arbitrary limitation in the application of provenance in archival practice is that archival descriptions of provenance tend to be parsimonious [17]. There are, of course, some extensive descriptions of the provenance of records [18]; however, even in such cases the description of provenance tends to be limited to unembellished basic factual information that provides a picture of the overall context in which records were created and kept, and it is often articulated at the

[1]The author was the NIH Records Management Officer from 1978 through 1988.

level of the person or organization that created the records, rather than the specific processes in which they were created. This level of description of provenance rarely offers any in-depth insight into the genesis of any individual record or aggregate of records below the level of the archival fonds.

A third aspect of the arbitrary narrowness of the concept of record provenance is that, in the digital realm it cannot be assumed that records are not changed once they are filed. Record provenance could logically be extended from origin to the entire life cycle of records to address changes that occur after creation in a comprehensive and coherent framework.

The concept of provenance also suffers from overextension. Professional literature over the last several decades has often argued for broadening the scope of both the concept and what is included in it. However, these arguments render the concept so broad and vague as to be extremely difficult to apply practically [19].

Like provenance, the concept of original order also has problems at both the conceptual and practical levels. Original order is fundamentally important because the way an actor organizes the information used in carrying out its activity supplements the information contained in the records. First of all, very few actions are carried out using a single document. Thus, the set of records kept in an activity indicates the range and variety of information the actor considered relevant and valuable in that activity. Second, the act of keeping records indicates at least an implicit judgment by the actor that the value of the records extended beyond their initial use. Third, the organization of the records reflects how they fit into the activities in which they were used or expected to be used.

Original order is closely linked to the existence of records because, obviously, to be a record a document must be kept. In essence the organization of records is a solution to a requirement; namely, readily accessing all of the information that is most likely to be relevant to a current action and only that information. In many cases, that requirement is best satisfied by grouping together records that were used in a prior action. This solution is effectively regarded as a requirement in records management standards [2, 20–23, 36], in government regulations [3], in policies of private sector and non-governmental organizations [24–26], and in archival theory [27]. The organization of records in filing systems is conceptually sound and demonstrably beneficial; nevertheless, it is an artifact of the technology of hard copy records and is limited by the fact that, with that technology the optimal, if not the only, effective and efficient way to arrange records is by physical proximity. The success of filing systems in practice does not justify an assertion that they are either the only or the best way to manage records. ICT opens new possibilities that might prove superior; for example, managing records on the basis of where and how often they are communicated.

The organization of records materializes relationships among records and these relationships can reflect how they were used by the recordkeeper, but this is not necessarily the case. One could assume that records organized in case files assemble the most significant records used in the activity that defines the case and arrays them in the sequence in which they were created, but this assumption is not valid for other ways of filing records, such as in subject files or correspondence files.

A subject does not necessarily correlate to an activity or set of related activities, and a given correspondent may communicate about a range of subjects or activities. Moreover, even when files are organized to correlate with activities, it is unlikely, except in the simplest cases, that all of the records that had a significant impact on a particular case are assembled in the case file [28]. Overemphasis on original order can lead to the assumption that the filing of records equates with the totality of relationships among those used in a given activity or to a mistaken assumption that it represents a complete and unbiased expression of the relationships between records and activities [29]. As Georgio Cencetti stressed in articulating the concept of the archival bond, the relationships among records that result from their use in the same activities are inherent in the records [30]. They may be reflected in the way records are aggregated, but they are prior to and independent of the organization of records. Even when it does reflect the archival bond, treating the aggregation of records in files as the equivalent or even the preferred expression of the bond between records and activities leaves us ignorant of other evidence of this bond.

Like the concept of provenance, original order has been the subject of critical scrutiny in recent decades. Criticisms have included that the conceptualization and implementation of original order has failed to distinguish physical collections from logical arrangements [31]; that the assumption that there is a single, static set of relationships among records flies in the face of their recontextualization over time, notably in the processes of archival management [32]; that records creators may not explicitly organize records; and that the way the concept has been articulated is ill suited to the records of individuals [33].

The provenance and original order of records are closely related. Provenance describes the origination of records and original order their organization. In practice, however, use of the two concepts is essentially independent. Descriptions of the provenance of records focus on their creators, rather than the records themselves; moreover, they are commonly articulated at the level of the entire archival fonds. Application of the principle of provenance to actual collections of records is guided by and expressed in the "principle of provenance," which dictates that fonds be kept separate, but this principle fails to impose or even indicate any practice that would elucidate the provenance of individual records or record aggregates below the level of fonds. Respect for the original order of records impacts the management of records preserved in archives and is a cornerstone for the development of finding aids for those records, but both archival functions are accomplished in the main with little more than a bow in the direction of provenance.

3 Out of the Morass

Any path out of the morass described in the previous section needs to incorporate valuable observations that have surfaced in criticisms of the traditional approaches to provenance and original order while concentrating on articulating concepts in a manner that can be readily translated into practice. An obvious way to do this would be to disentangle the multifaceted and not necessarily compatible insights that have been advanced in recent criticisms, reformulating the conceptual foundations of records management and archives to achieve clearer differentiation and easier implementation. The challenges posed by explosive growth and increasing diversification of digitally encoded records make it highly desirable that reformulation express the requirements for managing records in a manner that enables automated implementation, verification, and measurement.

The open-ended growth in both quantity and variety of born-digital records presents both substantial challenges and unprecedented opportunities for the management and preservation of records. Besides the quantitative challenge, novel intrinsic properties of digital records, including genres that do not and even cannot exist outside of the digital realm, and different ways of expressing and preserving their relationships compel rethinking traditional concepts and re-examining established methods. Digital records offer an important opportunity to enrich the concept and expand the use of provenance below the level of the archival fonds. While this would be impractical with hard copy records, provenance of individual and aggregate digital records can be captured from metadata generated automatically when digital records are created, revised and used; for example, using transmission data for email and audit trails on system use. Furthermore, ICT provides tools that could be used to capture automatically additional data related to provenance.

This reformulation in practice could be complemented at the theoretical level by adopting and adapting concepts, methods and tools from other disciplines. Propitious opportunities for enriching the theoretical constructs applied in managing records come from systemic functional linguistics, which focuses on language "that is doing some job in some context" [34]. This emphasis on the function of language and the concomitant recognition that function cannot be understood apart from context parallels the view of records as instruments and by-products of the conduct of affairs. More specifically, systemic functional linguistics offers a systematic approach to capturing and organizing different aspects of the provenance of records through the adoption of its differentiation of context into field, tenor and mode of discourse. Field of discourse refers to the action or interaction in which language is employed. Tenor of discourse refers to the parties who participate in the activity, their roles, relationships and relative status. Mode of discourse refers to the role both spoken and written language plays in each context, and addresses how it is expressed, how it is organized, and what it achieves. Activities, participants, and the modes and functions of expression are all crucial in understanding records. Adopting the specific constructs of field, tenor and mode of discourse as they have been articulated in systemic functional linguistics offers the opportunity for describing facets of the provenance of records in a clear and empirically verifiable manner.

Another discipline that can enhance the treatment of archival provenance is the branch of mathematics called graph theory. Graph theory offers suitable, quantitative methods of analysis and opens possibilities for the use of automated analytical and visualization techniques that are well suited to the objectives of records management. Graph theory can be applied to capture, but still distinguish, different aspects of provenance. In graph theory, a graph consists of a set of nodes connected by arcs, where the nodes represent things and the arcs relationships between nodes. Graphs are differentiated according to the types of things represented as nodes and the types of relationships between the nodes. The different aspects of context distinguished in systemic functional linguistics can be used to define corresponding archival graphs. In a graph of the field of discourse, the nodes could be either different activities or the various steps or stages of a single process and the arcs the transitions from one activity or step to another. The tenor of discourse could be graphed by identifying each person or organization involved in an activity as a node. Different graphs would result from selecting different types of relationships between parties; for example, one graph might depict interactions between parties as arcs, while another might indicate the relative status of individuals within an organization. Graphs of the mode of discourse would encompass records as nodes. The organization of records in a record-keeping system would depict the classification and placement of records aggregations as arcs. Another graph of mode could display derivation relationships among records revealing connections that would not surface in a graph or the arrangement of records. For example, many records are articulated in accordance with directives, but regulations, policy statements and other directives are usually not filed with the records representing the instances in which such directives and other policy documents are applied. Graph theory opens possibilities for extensive and precise description of relationships among activities, parties, and records through the superimposition of graphs of each of them [35]. System and human generated data could also be used to construct heterogeneous graphs showing the involvement of persons and organizations in activities and their participation in the generation and use of records.

Graphs of the different facets of the context of records creation, keeping and use would also benefit researchers who use records. Contextual graphs could help them to discover related records. Moreover, they could expand or extend such graphs by adding nodes and/or arcs that are of particular interest in their research, even extending the graphs to include outside parties that interact with the persons or organizations that create or keep records, such as correspondents, customers, and authorities.

To conclude, this chapter has reviewed the conceptualization of archival provenance and the related concepts of archival fonds and original order and the ways these concepts have guided practice. In doing so, it has identified problems entailed by the qualitative and imprecise conceptual approach traditionally applied to the issues involved in the management, use, and preservation of records, and suggested a multidisciplinary strategy to reformulate basic concepts, positing the possibility of applying systemic functional linguistics and graph theory as possibilities for finding a way out of the existing conceptual morass.

References

1. Horsman, P.: The last dance of the phoenix, or the de-discovery of the archival fonds. Archivaria **54**, 1–23 (2002)
2. International Organization for Standardization: ISO 23081-1 Information and Documentation – Records Management Processes – Metadata for Records: Principles. ISO, Geneva (2006)
3. National Archives and Records Service of South Africa: Rec. Manag. Policy Man. **7**, 15–20 (2007)
4. United Nations Secretariat: Secretary-General's Bulletin: Record-keeping and the Management of United Nations Archives. United Nations. ST/SGB/2007/5 (February 17, 2005), https://archives.un.org/sites/archives.un.org/files/ST_SGB_2007_5_eng.pdf
5. 44 United States Code § 3301(a)(1)
6. International Council on Archives – Committee on Descriptive Standards, "Glossary of Terms Associated with the General Rules," General International Standard Archival Description ISAD(G), 2nd edn. International Council on Archives, Ottawa (1999), http://www.icacds.org.uk/eng/ISAD(G).pdf
7. InterPARES 2: Dictionary, http://www.interpares.org/ip2/display_file.cfm?doc=ip2_dictionary.pdf&CFID=6354166&CFTOKEN=11363511
8. Pearce-Moses, R.: Glossary of Archival and Records Terminology. Society of American Archivists, Chicago (2005). http://www2.archivists.org/glossary/terms/o/original-order
9. Scott, P.J., Finlay, G.: Archives and administrative change: some methods and approaches (Part I). Arch. Manuscr. **7**(3), 115–127 (1978)
10. Scott, P.J., Finlay, G.: Archives and administrative change: some methods and approaches (Part 2). Arch. Manuscr. **7**(4), 151–165 (1979)
11. Scott, P.J., Finlay, G., Smith, C.: Archives and administrative change: some methods and approaches (Part 4). Arch. Manuscr. **8**(2), 51–69 (1980)
12. Scott, P.J.: Archives and administrative change: some methods and approaches (part 5). Arch. Manuscr. **9**(1), 3–18 (1981)
13. National Institutes of Health: Application Submission System & Interface for Submission Tracking (ASSIST) User Guide. System Version 2.15. Document Version 6.3.0 (October 16, 2015), https://era.nih.gov/files/ASSIST_user_guide.pdf
14. National Institutes of Health: NIH Grant Application Submission and Review. Useful Web Links, http://public.csr.nih.gov/aboutcsr/NewsAndPublications/Publications/Documents/ReivewWebLinks.pdf
15. National Institutes of Health: NIH Guide for Grants and Contracts, https://grants.nih.gov/grants/guide/index.html
16. Collins, F.: Opportunities for research and NIH. Science **327**, 36–37 (2010). https://www.genome.gov/Pages/Newsroom/Webcasts/2010ScienceReportersWorkshop/FCIntro_Opportunities-for-Research-and-NIH.pdf
17. National Archives and Records Administration (U.S.): Online Public Access, http://www.archives.gov/research/search/
18. Provincial Archives of Alberta, Canada: An Administrative History of the Government of Alberta 1905–2005 (2006), http://www.culture.alberta.ca/paa/archives/research/documents/contents.pdf
19. Douglas, J.: Origins: evolving ideas about the principle of provenance. In: Eastwood, T., MacNeil, H. (eds.) Currents of Archival Thinking, pp. 23–43. CA. Libraries Unlimited, Santa Barbara (2010)
20. International Organization for Standardization: ISO 15489: Information and Documentation - Records Management. ISO, Geneva (2001)
21. International Organization for Standardization: ISO 15801 Trustworthiness and Reliability of Records Stored Electronically. ISO, Geneva (2009)
22. Object Management Group: Records Management Services (RMS) Version 1.0 (2011), http://www.omg.org/spec/RMS/1.0/PDF/

23. U.S. Department of Defense: Electronic Records Management Software Applications Design Criteria Standard (2007), http://www.dtic.mil/whs/directives/corres/pdf/501502std.pdf
24. Bilotto, A., Guercio, M.: The management of corporate records in Italy: traditional practice and methods and digital environment. Rec. Manag. J. **13**(3), 136–146 (2003)
25. British Broadcasting Corporation: Records Management Policy, Version 1.4 (2010), http://www.bbc.co.uk/guidelines/dq/pdf/media/records_management_policy_v1.4.pdf
26. Chachage, B., Ngulube, P., Stilwell, C.: Developing a model corporate records management system for sustainability reporting: a case of the Iringa region in Tanzania. S. Afr. J. Inf. Manag. **8**(1), 1–17 (2006). http://www.sajim.co.za/index.php/SAJIM/article/viewFile/217/213
27. Boles, F.: Disrespecting original order. Am. Arch. **41**, 26–32 (1982)
28. Thibodeau, K.: Out of Bounds: Breaking the Chains of Original Order to Exploit the Potential of the Archival Bond. Association of Canadian Archivists, Annual Conference 2014. Victoria, British Columbia (2014), https://www.academia.edu/18470021/Out_of_Bounds_Breaking_the_Chains_of_Original_Order_to_Exploit_the_Potential_of_the_Archival_Bond
29. Duranti, L., Guercio, M.: Research issues in archival bond. In: Bearman, D., Barata, K., Trant, J. (eds.) Electronic Records Research 1997: Resource Materials. Archives & Museum Informatics, Pittsburgh (1998). http://www.archimuse.com/erecs97/s1-ld-mg.HTM#_jmp0
30. Cencetti, G.: Sull'archivio come "Universitas Rerum". Archivi **4**, 7–13 (1937)
31. Yeo, G.: The conceptual fonds and the physical collection. Archivaria **73**, 43–80 (2012). http://www.members-archivists.ca:8080/ojs2/index.php/archivaria/article/view/313/413
32. Macneil, H.: Archivalterity: rethinking original order. Archivaria **66**, 1–24 (2008)
33. Meehan, J.: Rethinking original order and personal records. Archivaria **70**, 27–44 (2010)
34. Halliday, M.A.K., Hasan, R.: Language, Context, and Text: Aspects of Language in a Social-Semiotic Perspective. Oxford University Press, Oxford (1989)
35. Lemieux, V.L.: Toward a "Third Order" archival interface: research notes on some theoretical and practical implications of visual explorations in the Canadian context of financial electronic records. Archivaria 78 (2014). http://journals.sfu.ca/archivar/index.php/archivaria/article/view/13493
36. Hilbert, M.: How much information is there in the "Information Society"? Significance **9**(4), 8–12 (2012). http://onlinelibrary.wiley.com/doi/10.1111/j.1740-9713.2012.00584.x/pdf
37. National Archives of Scotland: Records Management (July 8, 2013), http://www.nas.gov.uk/recordKeeping/recordsManagement.asp

Part III
Library and Information Science Perspectives

Provenance in Digital Libraries: Source, Context, Value and Trust

Lucie C. Burgess

Abstract Provenance is used in digital libraries to denote authorship, origination or creation, information integrity, rights to re-use and exploit digital content, discovery and linking of data, security, accountability and in the context of digital preservation. The concept is applied in an inter-disciplinary sense in the Bodleian Digital Library, alongside the use of W3C standard PROV, as a useful data modelling framework for the Oxford University Research Archive. The application of provenance in the metadata of digital libraries is discussed in terms of entities, agents, activities, locations, concepts and annotations. We consider research challenges associated with provenance in digital libraries, including potential extensions to PROV, crowd-sourcing, applications to new forms of data and determinations of trust.

Keywords Data model • Digital archives • Digital humanities • Digital libraries • Digital preservation • Linked data • Ontology • Provenance • Trust

1 Overview of Interest and Application of the Concept of Provenance

At the Bodleian Libraries, University of Oxford, we are interested in the concept of provenance as practitioners working in the digital library field. The Bodleian was founded in 1602 by Sir Thomas Bodley and is one of the largest academic libraries in the world, serving 11,000 academics, 22,000 students and in total around 70,000 registered readers. It manages over 12 million bibliographic records, more than 2 million high resolution images of the Bodleian's most valuable collections, over 450,000 digitized books, 30,000 highly structured texts, a growing collection of research datasets, and numerous websites and applications curated in a virtual infrastructure that approaches petabyte scale. Our work in the digital library field tends to enable and facilitate research projects on issues of mutual interest led by academic departments, both within and beyond Oxford, which we hope in many cases can transition from being projects to live services embedded in the Library.

L.C. Burgess (✉)
Bodleian Libraries, University of Oxford, Oxford, UK
e-mail: lucie.burgess@bodleian.ox.ac.uk

V.L. Lemieux (ed.), *Building Trust in Information*, Springer Proceedings in Business and Economics, DOI 10.1007/978-3-319-40226-0_5

Provenance has a wide definition in library applications, relating to notions of ownership (for example, the provenance of a manuscript within a collection), authorship (for example, the provenance of a book or article), origination or creation (for example, the provenance of scientific data arising from experiments) and many other types of contextual information. Provenance is a well understood area in art and digital libraries, where lineage, pedigree and source play a major role in understanding how things have been derived, and in determining a collection's authenticity and value [1].

"Provenance, also known as lineage, describes how an object came to be in its present state, and thus, it describes the evolution of the object over time" [1]. In the context of e-science, where provenance is an important concept in the reproducibility of research, the concept is often sub-defined in the literature into 'data provenance' (the origin and context of data and transformations through which they are derived) and 'workflow provenance' (the record of the entire history of the derivation of the final output of the workflow, including recording the hardware, software and instruments used in an experiment).

Provision of the provenance of information resources on the Web can be used as a basis for the assessment of information quality, improving the contextual information behind generation, transformation and integration of information on the Web [2]. There has been a huge growth in the use of linked data in recent years. In the world of linked data, provenance information such as authorship or ownership provides context which can be used to link resources to other resources on the web, using semantic technologies such as RDF (the Resource Description Framework) and ontologies which specifically describe provenance information such as the provenance ontology, a W3C standard: the Dublin Core bibliographic ontology; the FOAF ontology and others [3]. Therefore provenance and other contextual metadata can be used to play an important role in discovery, search and retrieval. Provenance also can denote notions of trust through association (for example, "I found this archival document in the Bodleian Library therefore I trust it to be an authentic source[1]"; "I downloaded this dataset from data.gov.uk, and I trust the website, therefore I trust the data").

At the Bodleian Libraries, the notion of provenance is an essential part of our ongoing efforts to construct a rich and informative contextual framework for the digital objects in our collections and their exposure to users and digital services on the Web. According to the Bodleian's Head of Research and Development, Neil Jefferies, "We consider a digital object to be a composite entity comprising data and metadata, which derives much of its intellectual meaning from its provenance". Provenance, and other contextual information for digital objects, is increasingly important in open research environments, wherein organizations create and publish

[1]There are some well-known forgeries in the special collections of academic libraries but if an item is known not to be authentic this tends to be noted in a catalogue record, where available. However, most archives contain large uncatalogued collections and a researcher should always ask themselves searching questions around the provenance and authenticity of sources in the course of their research.

sets of open research outputs on the Web that are used, re-used and transformed by others.

2 The Relevance of the Concept of Provenance to Digital Libraries

In this section, we give an overview of the relevance of the concept of provenance to digital libraries generally, with some examples specific to the Bodleian Libraries. In the realm of libraries, provenance is relevant to information integrity, rights to share, re-use and exploit digital content, discovery and linking of data, security and accountability and digital preservation.

The Bodleian's domain of operations covers collecting the breadth of digital research outputs of the University (journal articles, book chapters, PhD theses, conference proceedings and, increasingly, datasets underpinning publications) as well as the products of in-house digitization of library materials and born-digital archival deposits. With such a broad overview it becomes evident that there is pervasive need to capture, preserve and disseminate provenance.

The Bodleian Libraries has developed and maintains ORA, the Oxford University Research Archive [4], which aims to ensure that the research outputs of the University of Oxford are accessible for the long-term. Contextual and provenance information in ORA enables digital objects to be searched and retrieved more effectively, and plays a role in denoting the reliability, trust, credit, and rights to benefit from exploitation of the research outputs [5].

The fact that a large amount of data on the web is derived by replication, query processing, modification or merging raises concerns of information quality [6]. When moving to a linked data representation of information, we discover that most library metadata consists of unqualified assertions (such as catalogue entries recorded by curators or researchers) and, taken as a corpus, usually of variable quality and therefore of limited utility. Addition of the provenance (of assertions) to the ongoing creation of contextual metadata allows the introduction of notions of evidence and authority that potentially greatly enhances the value, utility and quality of the metadata—and thereby its trustworthiness. This has an additional effect in that it allows and encourages scholars to correct and enhance metadata (subject to requisite permissions) since the provenance information allows the annotations or assertions to be citable/attributable to a scholarly source.

Provenance of annotations and assertions is particularly important in crowd-sourced projects. For example, in the crowd context we interpret provenance as maintaining a record of the linked-data generated/maintained by the crowd and the process(es) involved [7]. The Bodleian Libraries has participated in crowdsourcing initiatives, such as "What's the Score" [8] through which users can transcribe the text of nineteenth century parlour music. We see potential in such initiatives for describing, annotating and interpreting our special collections, for which cataloguing using specialist curatorial expertise would be prohibitively expensive across the entire collection. In July 2015 we launched Digital.Bodleian (http://digital.bodleian.ox.ac.uk) in which users can tag and annotate images for public view.

The concept of provenance is also highly relevant in the field of digital preservation. In practice, it is generally assumed by users that file storage systems preserve digital bitstreams (binary files) accurately. That may not be the case, however, since bitstreams are known to degrade over time. Maintaining the authenticity (trustworthiness) and provenance (in this context, the history of creation, ownership, accesses and changes) of the preserved objects for the long term is of great importance, since users must be confident that the objects in the changed environment are authentic. To validate authenticity of a preserved data object, provenance is needed, i.e., the documented history of creation, ownership, accesses, and changes that have occurred over time for a given data object. Also a means is needed to guarantee that the data is whole and uncorrupted (its integrity has been maintained) [9]. In library terms, the digital preservation metadata standard PREMIS [10] is increasingly evolving towards a provenance model.

In the humanities, a large amount of scholarly effort goes into reconstructing the context and thus provenance of the artefacts that survive from the past. Projects led by researchers at the University of Oxford in which the Bodleian digital library has played, or is playing, an important role in development include:

- *Cultures of Knowledge, Networking the Republic of Letters, 1550–1750* [11]. This project is using digital methods to reassemble the correspondence networks of the early modern period in Europe. It is starting to use provenance frameworks to capture biographical and prosopographical information as the agent and activity-focused approach is an excellent fit.
- *Medieval Libraries of Great Britain* [12]. This project aims to bring together data on medieval books with the libraries that held them.
- *15th Century Book Trade* [13]. This project aims to reconstruct the provenance of 450,000 copies of early incunabula (books printed between 1450 and 1500) to address fundamental research questions relating to the introduction of printing in the West.

A research question relating to these projects is whether use of the W3C standard provenance data model and ontology PROV would provide a more structured and improved framework to the data structures underpinning these digital projects, all of which rely on incomplete, fragmented and in many cases lost records from the past. Many cultural heritage projects use the CIDOC-CRM standard ontology [14] as a basis for linked data models, which bears some resemblance to PROV.

Finally, as has been mentioned above, in the scientific disciplines, there is growing recognition that journal articles are insufficient when it comes to enabling the reproducibility of research, one of the cornerstones of the scientific method. This awareness led initially to the requirement for mechanisms to enable data publication, dissemination and discovery and, subsequently, the realization that data was of limited utility without methodological "metadata" alongside, couched in terms that are very similar to historical provenance. Examples of research projects focused on reproducibility in which provenance plays a key role are offered by ResearchObject.org [15] and myExperiment [16].

3 Conceptualizations and Enactments of Provenance in the Bodleian Digital Library

Provenance is defined at several levels in the Bodleian digital library. We capture and represent a broad range of digital objects which themselves represent the contextual framework for the conventional physical objects (if a physical manifestation exists). These entities are a generalization of those seen in such diverse frameworks as TEI (Text Encoding Initiative) [17], CIDOC-CRM [14] and schema.org [18]. We use the terms ‘entity’, ‘agents’ and ‘activities’ in broadly the same sense as in PROV.

- Provenance of physical artefacts—refers to activities and agents that describe, create or modify an artefact (entity) or its context. The provenance of a physical artefact is represented through digital metadata.
- Provenance of digital artefacts—activities and agents that create or modify an artefact (entity), its context or metadata. For a digital surrogate—the provenance of the physical item and the surrogate will begin to diverge at the moment of digitization or digitalization (such as transcription). A similar effect happens when a copy of a born-digital artefact is archived. Expressing these multiple histories is an ongoing challenge. For example, the point at which Alan Bennett’s diaries were transcribed by his secretary into a digital form (both the paper versions and digital versions are held by the Bodleian Libraries) allowed the potential for errors to be introduced, or notes to be (mis)interpreted; the digital artefact is no longer the same as the physical.
- Provenance of assertions—allows the capture and expression of uncertainty, evidence and differences of opinion in a knowledge model—an essential characteristic of scholarly discourse that has been hitherto rather poorly served by most commonly available mechanisms. The RDF (linked data) representation of knowledge tends to assume binary logic for assertions, i.e. one-to-one causal relations. The notion of provenance provides a mechanism based on events/activities that allows these two approaches to be made compatible.
- They can be classified in terms of:
  - Entities—such as collections, works and instances—framing the more traditional library objects such as books and archival documents, but also newer ones such as datasets or annotations
  - Agents—People, organizations and workflows (software, instruments, etc.) that can create or change digital objects
  - Activities/Events—the key provenance objects that embody the essential creation/modification/deletion of objects

- Locations—precise coordinates and geopolitical entities the define geographical scope for assertions
- Concepts—Classifications, vocabularies and ontologies that categorize and structure other objects
- Annotations—additional assertions about relationships/properties of objects but also textual descriptions of provenance that have been inherited from various sources.

The RDF/linked-data model provides us with a good generic mechanism for expressing this information which also, crucially, defines the method by which the mechanism can be extended to accommodate new information through OWL ontologies and RDF schema.

There are two specific examples of the relevance of provenance in the context of the Bodleian Digital Library that I would like to consider. The first of these is integration of the PROV ontology into the data model that underpins ORA. The storage of research outputs in an open repository such as ORA requires that the research materials be described with contextual information such as authorship and publication. This drives faceted search and retrieval and is also important for applications such as digital preservation, citation and reproducibility of research (e.g., of research datasets) [5].

The data model has been devised within the broader context of a data modelling initiative called CAMELOT [19] that has as its aim the integration of data across multiple digital resources owned or managed by the Bodleian Digital Library. This integration, in turn, has been designed to allow cross-search and retrieval across multiple, seemingly un-related, silos of data, important in the context of multi- and inter-disciplinary research. The ORA data model is a representation of various scholarly outputs (theses, journal articles and so on) and includes a representation of the contexts with which the scholarly outputs are associated, and by which they can be comprehensively described and understood. Through this data model, scholarly output is associated with provenance information such as people, organizations, time and location. PROV was used in the ORA data model due to its simple activity representation that could be used repeatedly to describe any type of activity, and therefore any type of context that was needed to describe a scholarly output [5]. It should be noted that although CAMELOT has been adopted for ORA, it has not been widely implemented to date in other Bodleian Digital Library services. A consideration in adopting any bespoke data model is the benefit and functionality this brings compared with the sustainability challenges of keeping the model up-to-date with the evolution of technology.

The second example is the *Digital Safe* digital archiving project, with which the Bodleian Digital Library team has been involved over the last two years. This project aimed to consider the user requirements, service model, technical specification, costs, business model and legal/compliance issues around a digital archiving and records management service for the collegiate university. Stakeholders included central administrators in Oxford University Administrative Services, the National Perinatal Epidemiology Unit (NPEU), Oxford University Archives, curators from the Bodleian's special collections with expertise in digital personal and organizational archives and archivists from a number of Colleges. The provenance requirements of *Digital Safe* were considered by users to be paramount because of the highly sensitive nature of the data, such as patient records or closed digital archives containing personal data or legally privileged information. In this context the concept of provenance relates to that of security and accountability as well as trust. For example, in addition to being sure that the data came from the source attributed, stakeholders wanted to make sure that it was secure, had not been accessed by unauthorized users or been tampered with in any way. While not yet using PROV, the notion of provenance is a core concept and driver in the design of the service.

4 Key Challenges in Establishing Provenance in the Bodleian Digital Library

Provenance information is readily available in a large number of cases but almost always not in a form immediately amenable to representation in PROV (for example, in the case of uncertain or incomplete time information).

In practice, because of the broad nature of the Bodleian Libraries collections, which contain vast quantities of analogue, digitized and born-digital material, provenance information is still drawn from a wide variety of sources. Example sources would include online web forms, spreadsheets, TEI [17] for manuscript and EAD [20] for archival encoded text records, data feeds from repositories (OAI-PMH [13] and other RESTful protocols) and databases via machine-to-machine APIs.

Where structured data exists, then a mapping can be constructed between the data and the ontology. With growing diversity of formats and the evolution of standards, such mappings require a significant maintenance overhead, because they have to be continually revisited and must evolve to keep pace with change. For example the metadata describing digital artefacts in ORA has been mapped to the Dublin Core ontology, which can be more easily exchanged via the OAI-PMH data exchange protocol used widely in institutional repositories.

There is a tendency for library metadata standards to become over-prescriptive—sacrificing fidelity of data to adherence to a data model. A key feature of linked-data RDF is that it permits extensible and flexible knowledge models. There is a careful balancing act required between these two approaches so that we have the necessary standardization to ensure interoperability between systems while providing the flexibility to express and capture the assertions that scholars wish to make.

Much historical provenance information is in prose form or hand/type-written manuscripts that require digitization and/or keying. In the short term, these can be accommodated as annotations (e.g., in cataloguing or discovery systems) in order to establish trust and identity. Prose form can provide limited utility in terms of discovery or analytics, however. Translating provenance information from human-readable formats to machine-readable formats can improve searchability and underpin analytics, although it is our experience at the Bodleian Libraries that digitization, data extraction and data mapping can be time-consuming and expensive.

Unfortunately, capturing machine-readable provenance information in digital libraries from analogue content (archival and manuscript material, incunabula, analogue audio-visual collections, many newspaper collections) is often extremely difficult in practice. Most 'special collections' in libraries do not lend themselves to automated metadata extraction, mainly because optical character recognition techniques used to produce machine-readable data from the image outputs of digitization do not produce reliable results from handwritten and early printed materials and because the semantic properties of written languages change considerably over time. Even if machine-readable data could be captured reliably from digitized images, this would not necessarily reveal details of authorship, origination, creation or other properties of provenance discussed in this chapter because such information may not be recorded in the material or in the descriptive metadata associated with it. Capturing the provenance of special collections often requires painstaking research, deduction, inference and the specialist expertise of researchers and curators. However, such provenance metadata may then be stored in another human-readable but non-machine readable format, again requiring time-consuming and expensive digitization, or in databases that lack interoperability using proprietary data formats or that are not easily discoverable by search engines. For example, the Bodleian Libraries' unique and diverse special collections extend to 25 km of shelf space of which it is estimated that less than 2 % has been digitized and approximately 10 % has machine-readable descriptions that *might* contain provenance information. Similar problems exist in the analogue collections of libraries, museums, archives and art galleries all over the world.

There are promising efforts to capture structured machine-readable provenance metadata for artworks in the Art Tracks project led by the Carnegie Museum of Art [21]. In this project, researchers are attempting to structure provenance data using a re-codification of the American Alliance of Museums cataloguing standard, so that curators, scholars and software developers can create visualizations that answer questions that otherwise would be difficult or impossible to answer without digital techniques. It may be possible for the tools developed by this project to be extended to archival and manuscript material, for which similar problems in capturing structured provenance metadata are experienced.

The case for capturing provenance information on a large, automated scale becomes more compelling for born-digital objects from which structured provenance metadata can be more readily extracted, particularly for research outputs (journal articles, conference proceedings, digital books, and e-theses), research data and research software. Discipline-based research communities (e.g. BioSharing.org [22] in computational biology) and organizations such as the Research Data Alliance and the Digital Curation Centre publish models and standards which encourage the capture, management and preservation of provenance information in order to facilitate re-use and reproducibility of scientific endeavor. It is for this reason that we have focused our efforts on capturing provenance information where we believe it can offer the greatest potential and the best value for money from library budgets.

5 Research Challenges Associated with Provenance

Recognition of the importance of provenance is by no means universal although it is growing [23]. Frequently, in the context of digital libraries, provenance is couched in terms of history or ownership rather than wider definitions of provenance relating to data or methodology. There is very little discussion in the literature of the application of provenance concepts to the metadata for traditional archives (i.e., who prepared the archival description and what sources they used in doing so). It would be interesting to compare the use of provenance concepts in linked data vocabularies and ontologies to PROV and to think about whether PROV needs extension to respond to different use cases or advances in technology. It would be useful to understand how PROV could be extended to cope with scenarios frequently associated with archival and museum objectives, such as incomplete, fragmentary and missing information which, perhaps through use of broader provenance concepts, could lead to new inferences and discoveries around the context and lineage of traditional library objects.

While PROV has been implemented in many scientific data contexts, to our knowledge it seems not to have been adopted widely in semantic models in the archives and records community. In digital libraries and inter-disciplinary research contexts there is a need for a generic framework that can encompass multiple content types and information resources to allow researchers to create cohesive and manageable personal collections in the course of their research. Recently there has been some discussion of provenance concepts typically applied in an e-science context to digital libraries containing a wide diversity of metadata (see for example [24, 25]). It would be interesting to explore these broader notions of provenance further and consider their potential uses and implementations in the digital library context. In such inter-disciplinary contexts it would also be useful to have a conceptual definition of provenance that spans all of the different disciplines and frameworks, as PROV aims to do from a semantic web perspective. It may also be useful to employ a broader definition of provenance that incorporates the physical/digital transition.

In library terms the integration of provenance information can represent a significant change in cataloguing behavior or metadata capture, requiring a more quantitative and scholarly approach but also admitting the possibility that others (particularly scholars in the field, or the public, or disparate data sources via the linked open data cloud) may contribute in whole or in part to records. There is some research being conducted into the utility of provenance metadata in crowd-sourcing applications, for example for data maintenance [7] or to denote levels of trust in crowd-sourced annotations [26] but it would be interesting to see this research extended to applications in use in digital libraries and archives, for example in crowd-sourced cataloguing projects, or in large-scale collaborative knowledge bases such as wikidata.org [27].

As mentioned earlier, provenance of information is crucial in deciding whether information is to be trusted. In that context, there is potential for the wide commercial application of provenance in the Internet of Things. 'Distributed ledgers' using blockchain technology (underlying cryptocurrencies such as Bitcoin) can provide new ways of assuring ownership and provenance for goods and intellectual property [28]. By way of a recent example, the UK-based start-up company Provenance.org [29] has developed a real-time data platform that "empowers brands to take steps toward greater transparency by tracing the origins and histories of products … with our technology you can easily gather and verify stories, keep them connected to physical things and embed them anywhere online." The diamond industry is beginning to implement a system called Everledger, also based on blockchain technology, which establishes a digital "passport" for each diamond. This records its provenance, travel, and transactions with a unique cryptographic "fingerprint". The need for the determination of trust will increase in importance with the growth of ubiquitous connected devices in the Internet of Things, which itself will give rise to new forms of data; this is an area where provenance information can play a key role and presents a fascinating research challenge.

Acknowledgements I would like to thank my colleague Neil Jefferies, Head of Research and Development at the Bodleian Libraries, for the thought-provoking discussions we have held on the notion of provenance, and for his time in helping me to prepare for the InterPares Trust Interdisciplinary Workshop on Provenance in May 2015.

References

1. Omitola, T., Gibbins, N., Shadbolt, N.: Provenance in Linked Data Integration. Future Internet Assembly, Ghent, Belgium, 16–17 December (2010)
2. Freitas, A., Knap, T., O'Riain, S., Curry, E.: W3P: building an OPM based provenance model for the Web. Futur. Gener. Comput. Syst. **27**(6), 766–774 (2011)
3. Hartig, O., Hartig, O.: Provenance information in the web of data. In: Proceedings of the Linked Data on the Web LDOW Workshop at WWW, vol. 39, no. 27, pp. 1–9 (2009)
4. Oxford University Research Archive (ORA): Oxford University Research Archive. http://ora.ox.ac.uk/
5. Jones, T.G., Burgess, L., Jefferies, N., Ranganathan, A., Rumsey, S.: Contextual and provenance metadata in the Oxford University Research Archive (ORA). In: Metadata and Semantics Research Volume 544 of the Series Communications in Computer and Information Science, pp. 274–285 (2015)
6. Hartig, O., Zhao, J.: Publishing and consuming provenance metadata on the web of linked data. Lect. Notes Comput. Sci. (including Subser. Lect. Notes Artif. Intell. Lect. Notes Bioinformatics), vol. 6378 LNCS, pp. 78–90 (2010)
7. Markovic, M., Edwards, P., Corsar, D., Pan, J.: The crowd and the web of linked data: a provenance perspective. Crowd Tech. Rep. SS, pp. 50–51 (2012)
8. Zooniverse.org: What's the score at the Bodleian. http://www.whats-the-score.org/ (2016)

9. Factor, M., Henis, E., Naor, D., Rabinovici-cohen, S., Reshef, P., Ronen, S.: Authenticity and Provenance in Long Term Digital Preservation: Modeling and Implementation in Preservation Aware Storage. Society, pp. 1–10 (2009)
10. Library of Congress: PREMIS: Preservation Metadata Maintenance Activity (Library of Congress). http://www.loc.gov/standards/premis/index.html (2016)
11. CulturesofKnowledge.org: Cultures of Knowledge: Networking the Republic of Letters, 1550–1750. [Online]. Available: http://www.culturesofknowledge.org/ (2016)
12. University of Oxford: Medieval Libraries of Great Britain. http://mlgb3.bodleian.ox.ac.uk/ (2016)
13. University of Oxford: 15th century book trade project. http://15cbooktrade.ox.ac.uk/project/ (2016)
14. CIDOC-CRM.org: CIDOC-CRM specification. http://www.cidoc-crm.org/ (2016)
15. ResearchObject.org: ResearchObject.org. http://www.researchobject.org/ (2016)
16. myExperiment.org: myExperiment. http://www.myexperiment.org/home (2016)
17. Text Encoding Initiative: TEI: Text Encoding Initiative. http://www.tei-c.org/index.xml (2015)
18. Schema.org: Schema.org. https://schema.org/ (2016)
19. Jones, T.G., Jefferies, N.: CAMELOT data model web page. http://camelot-dev.bodleian.ox.ac.uk/?page_id=20 (2016)
20. Library of Congress: EAD: Encoded Archival Description. https://www.loc.gov/ead/ (2016)
21. Berg-Fulton, T., Newbury, D., Snyder, T.: Art Tracks: Visualizing the stories and lifespan of an artwork. MW2015: Museums and the Web, 8–11 April. http://mw2015.museumsandtheweb.com/paper/art-tracks-visualizing-the-stories-and-lifespan-of-an-artwork/ (2015)
22. BioSharing.org: BioSharing.org. https://biosharing.org/ (2016)
23. Moreau, L.: The Foundations for Provenance on the Web, vol. 2 (2010)
24. Chawuthai, R., Wuwongse, V., Takeda, H.: A formal approach to the modelling of digital archives. Lect. Notes Comput. Sci. (including Subser. Lect. Notes Artif. Intell. Lect. Notes Bioinformatics), vol. 7634 LNCS, pp. 179–188 (2012)
25. Huang, A.W.-C., Chuang, T.: Relations for reusing (R4R) in a shared context: an exploration on research publications and cultural objects. In: the 4th International Workshop on Semantic Digital Archives (SDA), in conjunction with International Digital Libraries Conference (DL2014), London, 8–12th September (2014)
26. Huynh, T.D., Ebden, M., Venanzi, M., Ramchurn, S., Roberts, S., Moreau, L.: Interpretation of crowdsourced activities using provenance network analysis. In: First AAAI Conference on Human Computation and Crowdsourcing, pp. 78–85 (2013)
27. Wikidata.org: Wikidata.org. https://www.wikidata.org/wiki/Wikidata:Main_Page (2016)
28. UK Government Office for Science: Distributed Ledger Technology: Beyond Block Chain. https://www.gov.uk/government/uploads/system/uploads/attachment_data/file/492972/gs-16-1-distributed-ledger-technology.pdf (2016)
29. Provenance.org: Provenance.org. https://www.provenance.org/ (2016)

Conceptual Provenance in Indexing Languages

Joseph T. Tennis

Abstract This chapter discusses conceptual provenance, the phenomenon of conceptual change in indexing languages through the introduction of the author's previous work. Observations from looking at large classification schemes, like ontogeny, semantic gravity, and collocative integrity, are introduced. The chapter closes with a discussion of key challenges in the methodology and outlines future research.

Keywords Indexing • Information retrieval • Knowledge Organization • Ontogeny • Provenance

1 Introduction, Motivation, and Rationale

Indexing languages are tools used in the aid of information retrieval and sense-making [1]. They comprise schemes, thesauri, ontologies, and taxonomies. Contemporary notable examples include the category systems of Wikipedia, Library of Congress Subject Headings used by libraries around the world, and the Gene Ontology used by scientists to understand genomes of the world.

These tools are constructed at given point in time, and good tools are informed by the literature and users at that time [2, 3]. As more literature is added to the collection represented, and as users' needs change, so too do indexing languages. This causes a shift in structure and semantics in the indexing language [4–9]. For example, in the 1913 Dewey Decimal Classification (DDC), number 397 was the single address for GYPSIES, NOMADS, AND OUTCAST RACES defined as:

J.T. Tennis (✉)
Information School, University of Washington, Seattle, WA, USA
e-mail: jtennis@uw.edu

V.L. Lemieux (ed.), *Building Trust in Information*, Springer Proceedings in Business and Economics, DOI 10.1007/978-3-319-40226-0_6

> [p]eople without nationalities who do not coalesce with the ruling people among whom they live. This includes Gypsy language, which has no place in the linguistic groups of 400, as the Gypsy people have no place in the geographic divisions of history[1] [10].

Both the language and the people (their culture, customs, contemporary socio-political situations) are no longer at that address, and have not been since 1958. They are handled by different numbers from a different part of the classification scheme. This phenomenon, while rich with examples from Dewey because of its age, is not the only indexing language that changes. The Wikipedia category system is another example.

The Wikipedia category system is only 8 years old, but it has changed dramatically. Dbpedia has captured snapshots of the provenance of categories in the Wikipedia category system. From a preliminary data analysis of their data we can see that from 2008 to 2012 there has been nearly a 170 % increase in the number of categories and 200 % increase in the density of interconnections between those categories. The next step is to investigate the semantics of these changes [4].

Because change in indexing languages is a persistent phenomenon, and there is no commonly accepted design amelioration, it constitutes an important research area in the field of knowledge organization. We need to design for change, and we need to understand the phenomenon to have an informed design pattern [22].

2 Definitions, Key Concepts, and Conceptualization of Provenance

In the field of knowledge organization we talk about scheme change, instantiation, and ontogeny. We do not talk about provenance, though I am arguing here that these are all related concepts. If provenance is defined as the chronology of custody and context (in the physical world often signaled by physical location) of some material, then we can see how revisions of indexing languages could change the context of a concept. With the change in context, the concept may change its meaning, and it is the meaning of the concept, in relation to other concepts and the documents they index that we care about in knowledge organization.

When we trace the history of a concept through revisions of indexing languages, we are studying the concept's *ontogeny* [2]. Ontogeny is a term borrowed from biology. In biology it describes the maturation of an individual of a species. In the case of humans we start out with gill-like anatomy and something that resembles a tail. We lose these as we mature. That is part of our ontogeny. With the example above, GYPSIES, we can see that it too has changed from the earliest versions of the DDC up to today.

---

[1] 397 is between 396 WOMEN'S TREATMENT AND POSITION and 398 FOLKLORE, PROVERBS (s.l.).

We observe ontogeny by looking at *scheme change*. Observing change in print-based indexing languages is transparent. In that context we can compare one print *edition*, say the first edition of the DDC with the second, and so on. In the online environment we have a challenge with tracking changes insofar as those changes may or may not be related to a particular version of an indexing language. This is because instead of issuing particular editions, as we did in the print-only world, we can now change the *state* of an indexing language by changing a single term and not cast the indexing language as embodied in a new edition [3]. Both states and editions are versions of the indexing language, and we know the provenance or ontogeny of a concept through these two kinds of changes.

Finally, we can only identify a concept by its *instantiation*. That is, we can only attempt to identify the persistence or discontinuity of a concept through observations we make of the indexing language, and specifically its terms. So we know a concept's ontogeny through its instantiation in a particular version of an indexing language—and the relationship of that version to previous versions (if it is the latest version).

To date our work has offered a range of observations that can be used to begin to design ameliorations to the problems identified with scheme change. We have also begun to observe how information professionals have reacted to scheme change, and surfaced some methodological issues that must be addressed by researchers interested in conceptual provenance [11, 12].

First, we can see there are three general types of scheme change. *Structural change* is the movement of concepts from one location in the scheme to another. *Word-use change* does not move a concept, but adds, replaces, or takes away words that instantiate the concept. *Textual change* is the change in relationship between terms in the indexing language, and the extension of the texts represented by that term. For example, the texts classified under CIVIL ENGINEERING might be very different now compared to those in 1930.

These kinds of changes over time also affect the ability of the scheme to bring together texts that were published during the whole life of the scheme [13]. That is, it affects the integrity of the scheme's design requirements to *collocate* items discussing the same concept. By observing subject ontogeny we can begin to measure the *integrity* of a scheme [14, 15]. With these measurements we can begin to ask what is the threshold of deviation that we consider tolerable given scheme change [23].

With regard to how information professionals react to changes in schemes, we have observed that some decide to eschew the new or revised scheme in lieu of what term they currently have representing texts in their collection. In some cases we have observed catalogers keeping books on Anatomy in an outdated class. The rationale might be that similar books are already there and there is no option for reclassing, so to help the library user, they deviate from the scheme. We have called this *semantic gravity* in relation to the power of the old term (or class) [14].

Finally, there are some methodological questions the researcher must ask herself as she engages in the study of conceptual provenance. The most striking has to do with time and how we know the object to study i.e., the concept. Our philosophical stance and theoretical lens as researchers must be taken into account when we make a claim about the persistence of concepts through time, as well as how we slice time for the purpose of analysis. Do we only consider multiple editions or states or do we consider different implementations of the same scheme [16, 17].

3 Methods of Capturing and Representing Provenance

While there might be many methods of ontogenic analysis, I analyze indexing languages, the decisions made by indexers in relation to those indexing languages, and any and all contextual data that support a reading of those concepts in context (see, for example, [3, 18–20]). I have created charts that map concept relationships over time, and also where indexers agree or disagree with the indexing language (see Appendix). Because we do not often re-index collections, it is relatively easy to capture the indexer response, but it is not unambiguous. So there are methodological concerns related to the faithful identification of instances of concepts in particular versions of indexing languages, and across versions. This means we can capture some sense of the ontogeny (provenance), but it is always provisional and requires contextualization and an argument justifying the choice of interpretation [21].

4 Key Challenges in Establishing Provenance

As alluded to above, one key challenge is interpreting persistence or discontinuity in a concept over time. Is an ontogeny real or does it matter? These are key questions that have to be sorted out in relation to the purpose(s) of indexing languages, and the value added by indexing. Further, the date of indexing is a methodological concern, since it is not explicitly stated, but inferred from other data. While recently more data is being provided about date of indexing, this is not the case with older activities. So this too is a challenge for establishing the indexer reaction in relation to the concept ontology.

5 Future Research Challenges

As we expand our sample of indexing languages we will need to frame the problem and define types of indexing languages in relation to versions, instantiation, and ontogeny. Further there are different ways of conceptualizing time in relation to the concept ontogeny and the question, central to this workshop, provenance. Where did the concept come from? Has it changed? Can we track that change? What does it matter to the functioning of this particular (kind of) indexing language? What we need is more data, more wooly examples, and a greater range of arguments for or against this conceptualization of provenance.

6 Concluding Remarks

While it may be seen as a rarified endeavor to study the scheme change and subject ontogeny, i.e., conceptual provenance, it is of utility to the design of information systems as well as the forensic analysis of the semantics of large-scale long-lived systems. With knowledge of changes, we can build supports for our systems so that when they do change, they neither veer from their design requirements nor confuse users. Likewise, the investigation into conceptual provenance allows the researcher to see how decisions have affected the scheme. We learn about the semantic life of the scheme and in that context better understand the systems that we rely on to do our work.

Appendix

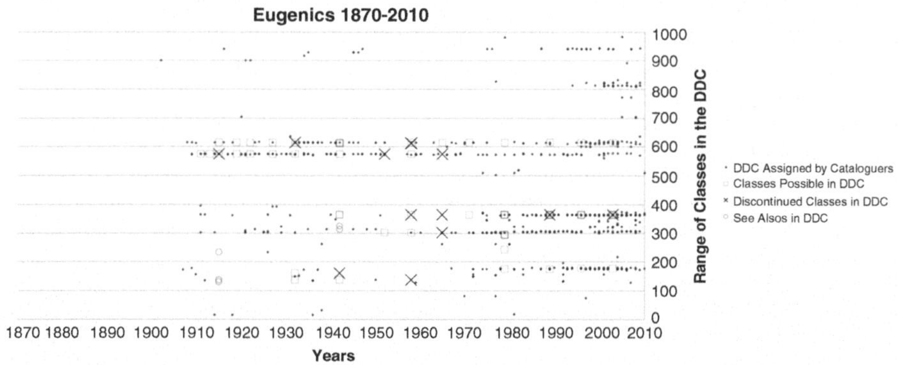

References

1. Dewey, M.: Decimal Classification and Relative Index for Libraries, Clippings, Notes, etc., 8th edn. Forest Press, Tionesta (1913)
2. Tennis, J.T.: Subject ontogeny: subject access through time and the dimensionality of classification. In: Challenges in Knowledge Representation and Organization for the 21st Century:

Integration of Knowledge across Boundaries: Proceedings of the Seventh International ISKO Conference, Granada, Spain, July 10–13, pp. 54–59. Ergon, Würzburg (2002)
3. Tennis, J.T., Sutton, S.A.: Extending the Simple Knowledge Organization System (SKOS) for concept management in vocabulary development applications. JASIST **59**(1), 25–37 (2008)
4. West, J., Tennis, J.T.: Categorical Landscapes: Large Scale Cluster Analysis of Wikipedia's Category System Over Time. iSchool Research Fair, Seattle, Washington (2012)
5. Tennis, J.T.: Emerging Concepts in Ontogenic Analysis. iSchool Research Fair, Seattle, Washington (2012)
6. Tennis, J.T., Thornton, K., Filer, A.: Some Temporal aspects of indexing and classification: toward a metrics for measuring scheme change. In: Proceedings of the 2012 iConference, pp. 311–316. ACM, NY (2012)
7. Thornton, K., Tennis, J.T.: Eugenics and Anatomy: Case Study of Change and Stasis in Classificatory Structures. iSchool Research Fair, Seattle, Washington (2010)
8. Tennis, J.T.: Facets and fugit tempus: considering time in faceted classification schemes. In: Proceedings of the 12th International Conference for Knowledge Organization, pp. 58–62. Ergon, Würzburg (2012)
9. Tennis, J.T.: A kaleidoscope perspective: change in the semantics and structure of facets and isolates in analytico-synthetic classification. SRELS: J. Inf. Manage. **50**(6), 789–794 (2013)
10. Tennis, J.T.: Foundational, first-order, and second-order classification theory. Knowl. Organ. **42**(4), 244–249 (2015)
11. Tennis, J.T.: Methodological Constructs in Subject Ontogeny Research. iSchool Research Fair, Seattle, Washington (2014)
12. Tennis, J.T.: Load bearing or Levittown? The edifice metaphor in conceptualizing the ethos of classification work. In: Proceedings of the 13th Annual Conference on Knowledge Organization, pp. 222–227 (2014)
13. Adler, M., Milojević, S., Rogers, C., Tennis, J.T., van Hooland, S., West, J.D.: The temporal dimension in the study of knowledge bases: approaches to understanding knowledge creation and representation over time. In: Proceedings of the ASIST Annual Meeting (2013)
14. Tennis, J.T.: The memory of what is: ontogenic analysis and its relationship to ontological concerns in knowledge organization. In: Smiraglia, R. (ed.) Ontology in Knowledge Organization. Ergon-Verlag, Würzburg (2015)
15. Tennis, J.T.: The strange case of eugenics: a subject's ontogeny in a long-lived classification scheme and the question of collocative integrity. JASIST **63**(7), 1350–1359 (2012)
16. Tennis, J.T.: Metaphors of time and installed knowledge organization systems: Ouroboros, Architectonics, or Lachesis? In: Proceedings of the Eighth International Conference on Conceptions of Library and Information Science, Copenhagen, Denmark, http://informationr.net/ir/18-3/colis/paperC38.html (2013)
17. Tennis, J.T.: Measured time: imposing a temporal metric to classificatory structures. In: Proceedings of the 11th International Conference for Knowledge Organization, pp. 223–228. Ergon, Wurzburg (2010)
18. Tennis, J.T.: SKOS and the ontogenesis of vocabularies. In: Proceedings of the International Conference on Dublin Core and Metadata Applications: Vocabularies in Practice, pp. 275–278. http://dcpapers.dublincore.org/pubs/issue/view/28 (2005)
19. Tennis, J.T.: Diachronic and synchronic indexing: modeling conceptual change in indexing languages. In: Information Sharing in a Fragmented World, Crossing Boundaries. Proceedings of the 35th Annual Meeting of the Canadian Association for Information Science/L'Association canadienne des sciences de l'information, pp. 1–11. http://dlist.sir.arizona.edu/1898/ (2007)
20. Tennis, J.T.: Edifice Complex: Conceptualizing Classification in 1890 and Now. iSchool Research Fair, Seattle, Washington (2013)
21. Tennis, J.T.: Scheme versioning in the semantic web. Cat. Classif. Q. **43**(4/3), 85–104 (2007)
22. Tennis, J.T.: Constructs in knowledge organization systems: rhythm in time, intention, and form. In: Contextes, langues et cultures dans l'organisation des connaissances: actes du 9e Congrès ISKO France 2013, 10 & 11 octobre, pp. 10–21. ISKO-France, Houdemont (2014)

23. Tennis, J.T.: Collocative integrity and our many varied subjects: what the metric of alignment between classification scheme and indexer tells us about Langridge's theory of indexing. In: Proceedings of the North American Symposium on Knowledge Organization. isko-cus.org/NASKO2013proceedings/Tennis_CollocativeIntegrity.pdf (2013)

Part IV
Computer Science Perspectives

A Brief Tour Through Provenance in Scientific Workflows and Databases

Bertram Ludäscher

Abstract Within computer science, the term *provenance* has multiple meanings, due to different motivations, perspectives, and assumptions prevalent in the respective communities. This chapter provides a high-level "sightseeing tour" of some of those different notions and uses of provenance in scientific workflows and databases.

Keywords Lineage • Prospective provenance • Provenance games • Provenance polynomials • Retrospective provenance • Why-not provenance

1 Introduction: Provenance in Art, Science, Computation

The Oxford English Dictionary (OED) defines provenance as "the place of origin or earliest known history of something; the beginning of something's existence; something's origin." Another meaning listed in the OED is "a record of ownership of a work of art or an antique, used as a guide to authenticity or quality."

In the fine arts, the importance of this notion of provenance can often be measured with hard cash. For example, one of Picasso's *Les Femmes d'Alger* sold for nearly \$180 million in May 2015 at Christie's in New York; a new record for a painting at an auction. In contrast, *La Bella Principessa* sold for less than \$20,000 in 2007, despite the fact that some attribute it to the great Leonardo da Vinci (Fig. 1a). However, there is no documented *chain of custody* prior to the twentieth century, so the drawing's incomplete provenance record is insufficient to establish its authenticity. It is now up to "provenance sleuths" to try and determine whether or not the drawing was really created by da Vinci—in which case it could rival the value of *Les Femmes d'Alger*.

Scientists often have to be expert provenance sleuths themselves. As part of conducting their science they may, e.g., analyse the stratigraphy of the Grand Canyon in order to reveal the geologic history of the planet (Fig. 1b), or study the fossil record preserved in rock layers or the molecular record inscribed in the

B. Ludäscher (✉)
School of Information Sciences and National Center for Supercomputing Applications, University of Illinois at Urbana-Champaign, 501 E Daniel St, Champaign, IL, USA
e-mail: ludaesch@illinois.edu

V.L. Lemieux (ed.), *Building Trust in Information*, Springer Proceedings in Business and Economics, DOI 10.1007/978-3-319-40226-0_7

Fig. 1 Provenance in the Arts and Sciences: (**a**) *La Bella Principessa*, portrait by Leonardo da Vinci. Or is it? It could be worth well over $100 million dollars, *if* enough provenance were available to verify its authenticity. (**b**) Grand Canyon's rock layers are a record of the early geologic history of North America. The ancestral puebloan granaries at Nankoweap Creek tell archaeologists about the much more recent human history. (By Drenaline, licensed under CC BY-SA 3.0)

DNA of species to reconstruct phylogenies and assemble the tree of life. Empirical evidence plays a crucial role in the scientific method and is a form of provenance that is everywhere around us, from the cosmic microwave background left behind by the Big Bang, to the recurrent laryngeal nerve we share with all tetrapods [1]—clear evidence of our common lineage with all life [2].

1.1 Transparency and Reproducibility in Science

It is long standing practice to cite your sources in scientific publications. However, as science has become increasingly computational and data-driven [3], and more interdisciplinary and collaborative, new requirements and opportunities have emerged for research articles. The U.S. Global Change Research Program (USGCRP) has developed the Global Change Information System (GCIS) [4] that links global change information across many federal agencies. An important product of USGCRP is the National Climate Assessment (NCA) report [5] which summarizes impacts of climate change on the U.S., now and in the future. To facilitate transparency and usability of the NCA, ambitious transparency goals have been set, ranging from basic source traceability (references to papers) to the use of data citations and metadata, all the way to traceable processes and software tools, with the ultimate goal to support full reproducibility of all NCA content [6].

Data provenance, the lineage and processing history of data, is of critical importance for transparency, to assess data quality [7], and for computational

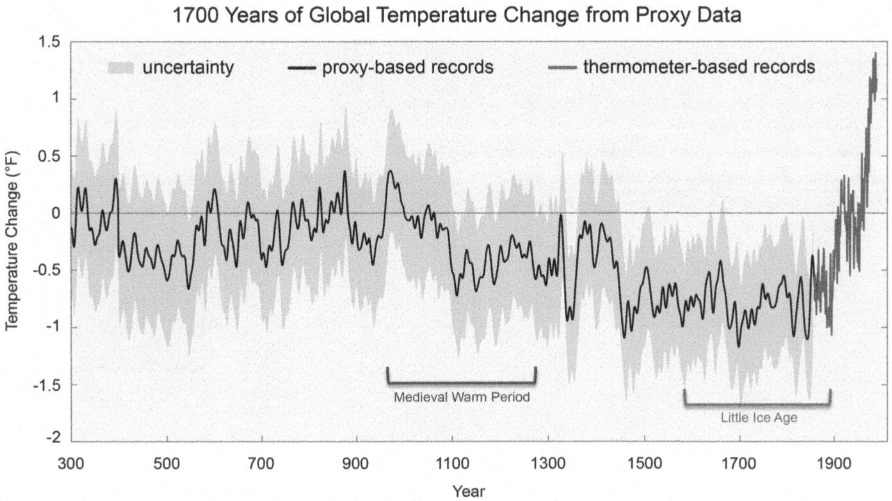

Fig. 2 "Hockey stick" graph from [5] (adapted in turn from [8]) showing temperature changes of the Northern Hemisphere from observations (*red*) and proxies (*black*) relative to the 1961–1990 average temperature (*gray* 0 °F line) (Color figure online)

reproducibility. Consider, e.g., the famous "hockey stick" graph in Fig. 2, showing temperature changes over the last 1700 years. Similar to *La Bella Principessa*, the value of such a chart may depend on its provenance, in particular, on the quality of the data that went into it, and the soundness of the computational method used to create the final result. As scientists provide detailed provenance information, e.g., *what* proxy records where used to reconstruct past temperature data and *how* those proxies where processed to derive a temperature, other scientists can evaluate and assess the results and the validity of the findings.

In a recent article, Hill et al. [9] make a strong case for data provenance for science. They cite a study by Eisenman et al. [10] that argues that the Antarctic sea ice extent was probably not growing nearly as fast as thought, and that "much of this [ice] expansion may be a spurious artifact of an error in the processing of the satellite observations." Hill et al. also report that ESIP[1] seeks to accelerate the implementation of new approaches to track all details necessary to demonstrate data validity and to ensure scientific reproducibility using a Provenance and Context Content Standard (PCCS) [9].

The third NCA report provides some of the much needed provenance and context information through the related GCIS system. Figure 3 depicts a screenshot showing rainfall *vs* temperature data. Metadata provided for the scatter plot in the upper right of the figure includes its spatial extent (lower right) and its temporal extent (the years from 1895 to 2012). Last not least, provenance links to the original dataset

[1]The Federation of Earth Science Information Partners.

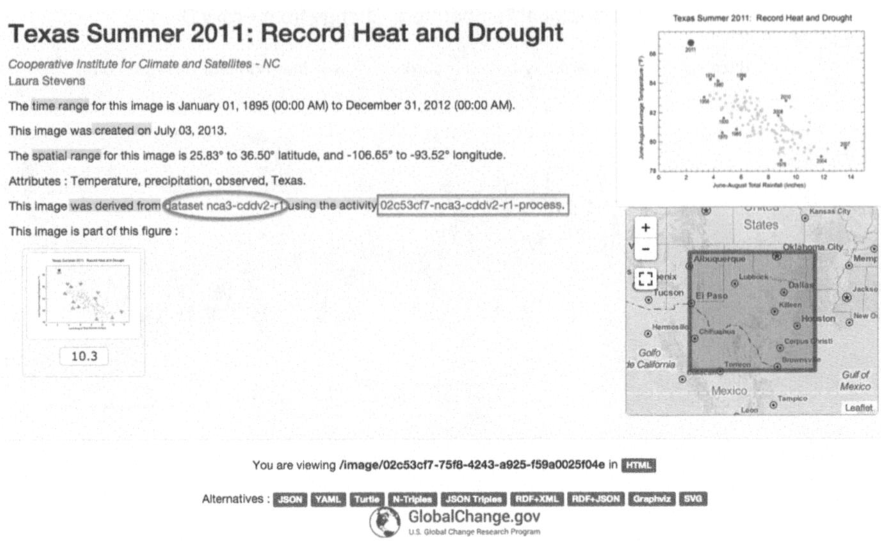

Fig. 3 Screenshot from data.globalchange.gov, showing a rainfall *vs* temperature scatter plot for Texas between 1895 and 2012 (*upper right*); provenance metadata (*center*) with links to the source data (*highlighted oval*) and software (*highlighted rectangle*) used to create the plot [11]

and software are highlighted in this HTML metadata view as well. By pushing one of the buttons at the bottom of the screen, this metadata can also be exposed in one of several other machine-readable formats, including JSON, YAML, Turtle, and RDF. While this rich metadata and provenance information is clearly useful and required for transparency, the compilation of this information for the report and the GCIS system required an extraordinary 3-year effort by a team of more than 300 experts [5]. As more and more workflow tools and scripting environments become "provenance-enabled", the capture, sharing, and querying of provenance information in support of reproducible science should become easier as well.

2 Provenance in Scientific Workflows

A scientific workflow is a description of a process for accomplishing a scientific objective, usually expressed in terms of tasks and their dependencies [12]. Such workflows aim to support computational science and accelerate scientific discovery in various ways, e.g., by providing process ***A**utomation*, ***S**calable execution*, ***A**bstraction*, and ***P**rovenance* support (ASAP for short) [13]. The latter, i.e., the automated tracking of provenance is often considered one of the key advantages of using a workflow system for process automation [14, 15].

Common processing examples include data formatting, subsetting, cleaning, and analysis. Compute-intensive workflows often result from computational science *simulations*, e.g., running climate and ocean models, or other simulations from particle-physics, chemistry, biology, to ecology, astronomy, and cosmology [16]. Scientific workflows can be simple, linear chains of tasks, but more complex dataflow graphs are also common [17].

2.1 Workflows as Prospective Provenance

Figure 4 depicts an example scientific workflow for the semi-automatic curation of specimen collections data [18], implemented using the Kepler scientific workflow system [19]. In Kepler, computational steps execute independently from one another and are implemented by so-called (software) *actors* (green boxes in Fig. 4). These actors are connected via dataflow *channels* that are typically implemented using FIFO (first-in first-out) buffers, i.e., in such workflows data elements can be executed in pipeline-parallel mode, similar to the way a UNIX pipeline executes. The workflow in Fig. 4 reads as input a CSV file containing specimen records from a natural history collection. Such biodiversity datasets may require time-consuming, manual data curation steps. Using workflow tools, a number of data quality control measures and repair suggestions can be processed more efficiently. The curation workflow in Fig. 4 checks various fields of the data records as they are streamed through the process pipeline, e.g., the plausibility of geolocation information (where a specimen was collected), the scientific name of the specimen, and the flowering time (for plants an additional check on the collection date). Further downstream, human actors are involved in checking the records flagged by upstream computational steps [18]. The final steps of the workflow display record locations on

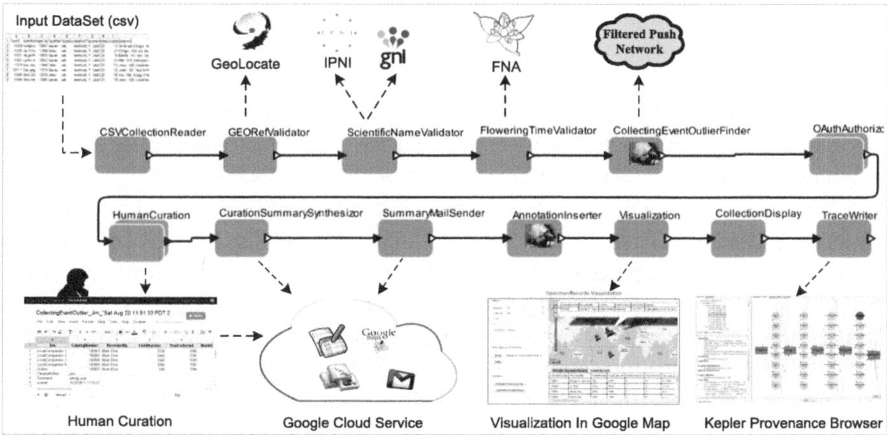

Fig. 4 Kepler data curation workflow for specimen data [18]. The workflow graph itself represents *prospective* provenance. The trace graph (*retrospective* provenance) depicted in the lower right can be viewed with a separate application; see Fig. 5

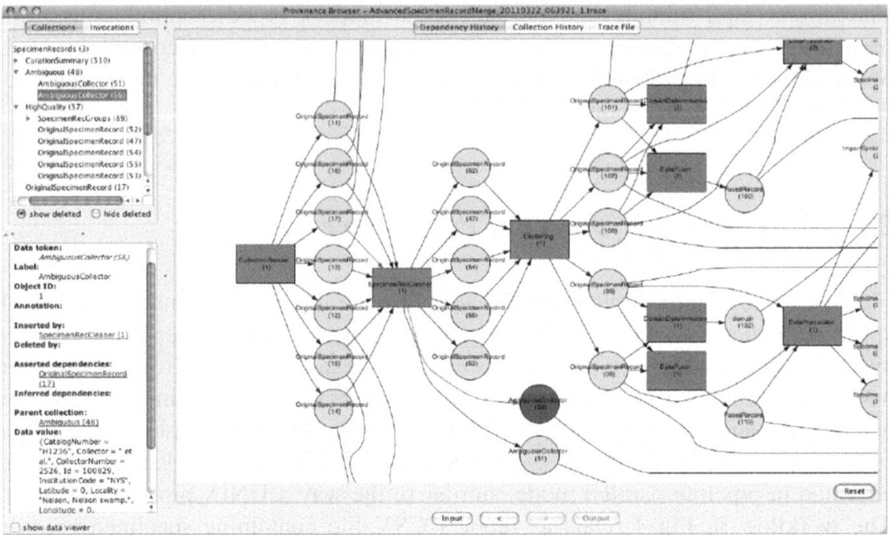

Fig. 5 Kepler Provenance Browser [20, 21]: A *retrospective provenance* graph (recorded earlier, during workflow execution) is displayed and can be navigated forward and backward in time via VCR-like control buttons (*bottom*)

a map and output a *provenance graph* that can be queried and explored in a separate provenance browser [20, 21].

The curation workflow graph depicted in Fig. 4 provides an overall description of the processing steps that a data record will undergo when subjected to the workflow. In this way, workflows are a form of *prospective provenance*: the workflow graph captures the general method or "recipe" of how data products of a workflow are processed. When a computational method is documented in this way, as a workflow graph, users can already make certain inferences about the general method and about the result data produced by it. For example, from the graph in Fig. 4 we see that the flowering time validation step (FNA) may use the improved geolocation data (GeoLocate) or a validated scientific name (IPNI/gni) since those upstream actors may have updated a record by the time it reaches the FNA step. Conversely, as the FNA actor lies downstream from GeoLocate and IPNI/gni, it *cannot* possibly influence the latter. Thus, while detailed dependency and lineage information between concrete data products is available only after workflow execution, some lineage information, in particular about the *independence* of steps can be obtained prior to execution, by querying the workflow graph. If a workflow graph contains further configuration information, e.g., which XML elements of a data stream are processed at each step, then a more detailed prospective provenance graph can be inferred as well [22].

2.2 *Retrospective Provenance from Workflow Execution Traces*

Prospective provenance, in the form of a workflow graph, constitutes a first valuable knowledge artifact, documenting a computational method or workflow. Many workflow systems also allow users to record provenance information at runtime, i.e., they capture *retrospective provenance* that can be queried, analyzed, and visualized to gain a deeper understanding of how certain results were obtained as the workflow executed. Figure 5 depicts a screenshot of the Kepler Provenance Browser [20, 21], showing retrospective provenance from a run of a specimen curation workflow similar to the one in Fig. 4. Selected nodes and incident edges are highlighted to indicate which upstream step has generated a data item, and which downstream step(s) read it. Note that a single actor in a prospective provenance graph can give rise to multiple *invocations* in the retrospective provenance graph, e.g., `DataFuser(1)` and `DataFuser(2)` in Fig. 5 are two distinct invocations of a single `DataFuser` actor. Each invocation usually operates on its own data items (beige circles). Similarly, a single channel between connected actors in the workflow graph (prospective provenance) is often traversed by multiple data items which then appear as "data bundles" in the execution trace (retrospective provenance graph), as seen in Fig. 5.

2.3 *Models of Provenance and Scientific Workflows*

In 2006 the scientific workflow community organized the first "Provenance Challenge" workshop to better understand the capabilities of different workflow systems and approaches [23]. The first workshop led to a number of follow-up challenge events (all set up to be informative rather than competitive), ultimately leading to the definition of the Open Provenance Model (OPM) [24, 25], which in turn informed the development of the W3C PROV standard [26]. Much work in the scientific workflow community then focused on engineering challenges, e.g., the efficient storage [27–29], navigation [30], and querying [31, 32] of provenance. When working with provenance in scientific workflows, the distinction between prospective and retrospective provenance is important. However, neither OPM nor its PROV successor deal with this distinction. One could argue that both OPM and PROV focus on retrospective provenance, but the underlying definitions are rather vague on that point.[2] As a result, different extensions to OPM and PROV have been developed that allow users to work with both prospective and retrospective provenance and relate both kinds of information in a single model [33, 34].

[2]For example, [26] states that "provenance is defined as a record that describes the people, institutions, entities, and activities involved in producing, influencing, or delivering a piece of data or a thing."

2.4 Relating Retrospective and Prospective Provenance

It is often desirable to combine trace-level retrospective provenance and workflow-level prospective provenance in a single, uniform representation. Such a model should also accommodate temporal information whenever available. This can be achieved with a semistructured data model, consisting of labeled, directed graphs of the form $G = (V, E, L)$, with vertices V, labels L, and labeled edges $E \subseteq V \times L \times V$. In the following, we view workflows (prospective provenance) W and traces (retrospective provenance) T as subgraphs of G. Similarly, a temporal model consists of labeled edges, modeling one or more "before" relations $\leq_R$.

Figure 6 shows a workflow W (top) and a trace T (bottom). By linking a trace to the workflow that generated it, important information can be obtained via the constraints of the combined model: If data item y is written into output container Y as a result of invocation a of actor A on input item x in X, then the writing of y cannot happen before x is read. Therefore, this *firing constraint* at the level of the workflow model W induces a corresponding *temporal constraint* on the trace T, i.e., $t_{\text{read}(x)} \leq_{\mathsf{f}} t_{\text{write}(y)}$. Similarly, the *data constraint* at the Y container in W induces another temporal constraint at the trace level: before item y can be read by invocation b of actor B, this item must first have been written by some invocation a of A, i.e., $t_{\text{write}(y)} \leq_{\mathsf{d}} t_{\text{read}(y)}$.

In [36] the authors use temporal information about the duration of interactions to exclude data dependencies that would violate temporal causality (if process A first writes y, then reads x, then y does not depend on x).

Structural and Temporal Constraints The execution of workflow W in Fig. 7a might have produced the trace T in Fig. 7b. To check whether T is indeed a possible instance of W, we link T's nodes and edges to W via a mapping h (as in Fig. 6). For example, the edges $x \overset{\text{read}}{\rightarrow} a$ and $a \overset{\text{write}}{\rightarrow} y$ in T (x was read and y was written by

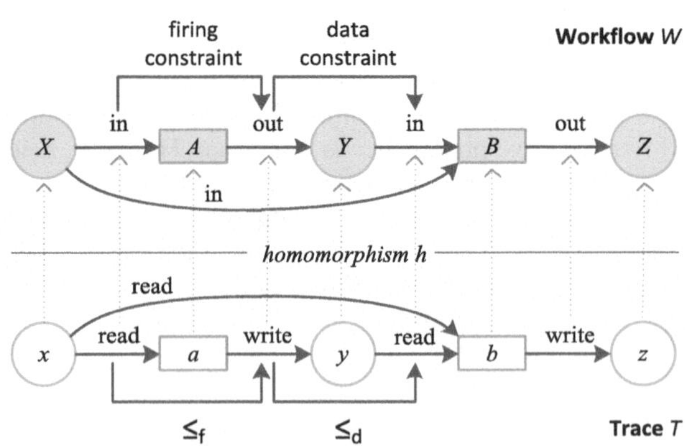

Fig. 6 A homomorphism h from trace T to workflow W guarantees structural validity. Workflow-level constraints induce temporal constraints $\leq_{\mathsf{f}}$ and $\leq_{\mathsf{d}}$ on traces [35]

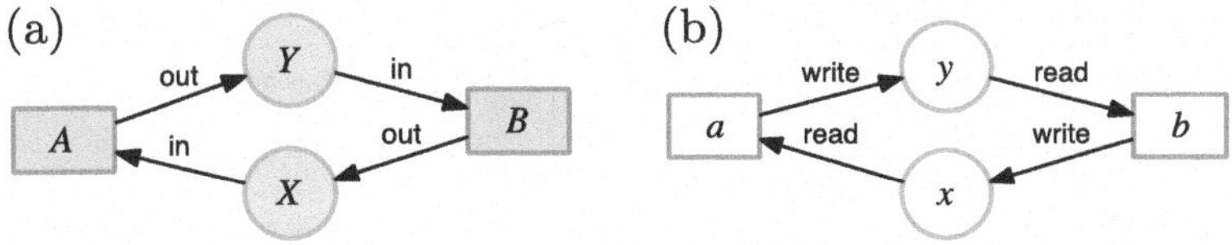

Fig. 7 Cycles (**a**) in workflow W and (**b**) in trace T. A cycle (feedback loop) in a workflow is not uncommon, but in a trace it suggests a temporal inconsistency [35]

invocation a), have corresponding edges $X \xrightarrow{\text{in}} A$ and $A \xrightarrow{\text{out}} Y$ in the workflow W, linking data containers X and Y to the actor A. In Fig. 7, T is structurally valid with respect to W, but other inconsistencies due to temporal constraints can still arise. For example, a cycle in T usually indicates an inconsistent trace: if read and write observables are temporally or causally linked, a *strict* partial order is implied and a cycle should not be observable. On the other hand, a cycle in W is usually *not* a concern. It simply means that W has a feedback loop, which is a rather common workflow pattern: loops in W are "unrolled" in T, leading to acyclic trace graphs T. In [35] we have formalized structural validity of a trace T via a homomorphism $h : T \to W$ and shown that it can be checked using a simple Datalog query.[3] In [37] a formal, temporal semantics of OPM is developed and it is shown that the original inference rules for OPM are sound but incomplete. In [38] we have developed a rule-based implementation (inspired by [37][4]) that allows provenance model engineers to experiment with different temporal semantics, expressed as constraints over the provenance model.

Example: Hamming Numbers Consider the two variant workflows H_1 and H_3 in Fig. 8a, b that compute *Hamming numbers*[5] [39, 40]

$$H = \{2^i \cdot 3^j \cdot 5^k \mid i, j, k \geq 0\}$$

incrementally, i.e., as an ordered sequence $1, 2, 3, 4, 5, 6, 8, 9, 10, 12, 15, \ldots$ While both workflows contain the same nodes (i.e., *actors* and *data containers*), they are wired slightly differently, which makes a big difference as it turns out. The data containers Q_i are queues (FIFO buffers); Q_8 is the distinguished output, where the Hamming numbers will appear in the correct order. M_1 and M_2 are *merge actors*, i.e., processes which take two ordered input sequences and merge them into an ordered output sequence. If presented with the same item in both streams, the output stream will only contain one copy of the element, so duplicates are removed. The actors

[3] Here, we are not *searching* for a graph homomorphism, but simply test whether the *given* mapping $h : T \to W$ is a homomorphism.

[4] ... or rather an earlier version from 2010: our 2013 paper could not have been influenced by a 2015 paper, nicely illustrating the very point of temporal constraints.

[5] Also known as *regular numbers*.

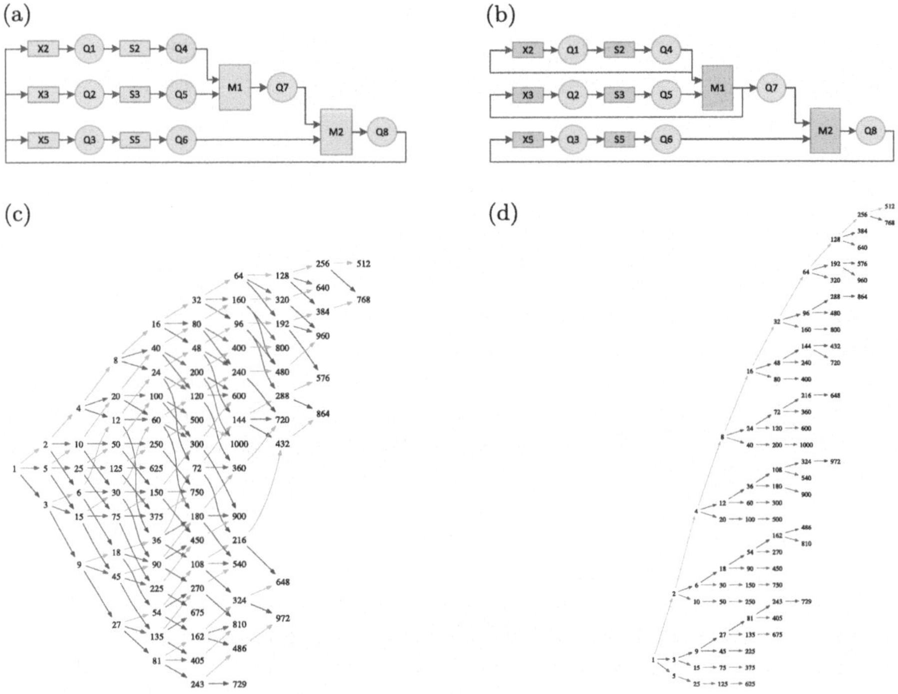

Fig. 8 Hamming workflow variants (**a**) H_1 ("one loop") and (**b**) H_3 ("three loops"). Retrospective provenance can be used to spot inefficient, redundant workflow computations: (**c**) trace T_1 ("*Fish*") obtained by running H_1 and (**d**) trace T_3 ("*Sail*") from H_3. The many redundant lineage paths of the DAG in (**c**) match the regular path query (x2 | x3 | x5)*, while the unique paths in the tree (**d**) satisfy the pattern (x2* · x3*) · x5*

x2, x3, and x5 multiply their inputs with 2, 3, and 5, respectively. Last not least, the *sample-delay actors* s2, s3, s5 are used "to prime the pump": initially (i.e., before reading any input), they output the number 1 to get the loops started. Subsequently, they simply output whatever they receive as an input. By design, the Hamming workflows H_1 and H_3 define an infinite output stream, i.e., these processes can "run forever".

Figure 8 shows two provenance traces T_1 (*Fish*) and T_3 (*Sail*) for Hamming numbers $n \leq 1000$, corresponding to the workflow variants H_1 and H_3. To save space, the trace graphs show each invocation of a multiplication actor x2, x3, and x5 as a colored edge (green, blue, and red, respectively). By querying the trace graph, the answer relation can be obtained as a set of edges $d_1 \xrightarrow{p} d_2$, linking data items to each other, with the (implicit) label p denoting the actor invocations (multiplication factors) used. Note that while the workflow graphs in Fig. 8 are cyclic, as expected, the trace graphs are acyclic. The trace-level retrospective provenance yields valuable information: In Fig. 8c Hamming numbers n can be produced in many different ways (if n contains all three factors 2, 3, and 5, its in-degree is always three). As a result, the provenance graph T_1 is not a tree, but a DAG (directed acyclic graph). In contrast, in Fig. 8d every Hamming number n is produced in one way only (there is a unique path from 1 to n), i.e., *without* creating unnecessary duplicates. Thus, unlike T_1, trace T_3 is a tree.

This example demonstrates another use of relating retrospective and prospective provenance, i.e., differences in the trace graphs T_1, T_3 can be used to explain the performance differences of the workflows H_1 and H_3 that generated them. Similarly, [41] uses retrospective provenance to compare the efficiency of different variants of a transitive closure query. Other works making use of the relationships between prospective and retrospective provenance include [42–46].

3 Provenance in Databases

When comparing data provenance in workflows and in databases, the former is usually considered a form of *coarse-grained* provenance, while the latter is considered *fine-grained* provenance. Indeed, provenance from workflows often captures observables at the level of files read and written by workflows or scripts [46]. In contrast, provenance in databases aims to answer record-level questions, e.g., which tuples (rows) in the input tables contributed to a particular output tuple and how [47]. Along another dimension, workflow provenance is sometimes called *black-box* provenance, whereas database provenance is considered *white-box* provenance [15, 48, 49]. This distinction is motivated by the fact that in workflows, the computational steps or actors are usually considered "black boxes" whose inner workings are not accessible or not relevant.[6] Conversely, as we shall see below, a database query can be considered a "white box", since its inner workings are readily available and analyzable [47, 53]. There are also approaches that combine workflow and database provenance, e.g., [54].

Database Provenance Questions In the following, we consider the most widely-used and best studied database model, i.e., the relational model [55]. But the basic principles usually also apply, *mutatis mutandis*, to other database models and queries, e.g., over semistructured (XML) data.[7] Consider a query answer $A = Q(D)$, i.e., an output table A resulting from the evaluation of a query Q on an input database D. Let $t \in A$ be a result tuple from the answer. In a database context, we would like to answer provenance questions such as:

What is the *lineage* of t, i.e., which specific subset(s) of tuples from the input D were used to produce t? Similarly, we might want to know *why* t is in the result and *how* exactly t was obtained from the tuples in its lineage. The notions of *lineage*,

[6] However, workflow systems such as Kepler [19] support nested workflows, so it is possible to open these "grey boxes" [14, 50, 51]. Similarly, fine-grained provenance from script-based workflows can be captured via profiling tools [52].

[7] For example, [56, 57] show how XML queries can be reduced to relational queries.

why-, and *how-provenance* (among others) have been formalized, studied in detail, and compared thoroughly [47]. Before we illustrate these different notions with a running example, we first give a brief (and necessarily incomplete) overview of some key publications and milestones in database provenance.

3.1 A Brief History of Database Provenance

The idea of propagating annotations from sources through queries to results is at the core of many current database provenance approaches, but also had early precursors such as [58], which proposed a model to carry along source attributions through queries. Another early approach which does not rely on annotations is described in [59]. Database research on provenance became mainstream through important, workflow-like applications in data warehouses [60]. Data warehouses periodically retrieve and integrate information from multiple sources using extract-transform-load (ETL) scripts, and then make the integrated information readily available for online analytical processing (OLAP) [61]. In data warehousing and other information integration scenarios, it is often crucial to be able to trace the lineage of data from output tables back to the sources where the data originated. In this way, data quality problems can be detected, localized, and eventually resolved.

An influential paper by Buneman et al. [62] developed the *why-provenance* model, refining another influential model by Cui et al. [60] for tracing lineage in data warehousing applications. The *provenance semiring*[8] framework developed by Green et al. [63] (and applied in a data sharing and information integration context [64]) marks a milestone in provenance research, as it subsumes many earlier provenance models and embeds them in a single, unified framework.

All provenance models mentioned so far aim at explaining, at various levels of detail, *why* and *how* a query answer $t \in Q(D)$ came about. Thus, these database approaches aim to relate outputs back to the inputs on which they depend, i.e., at a high level, they resemble retrospective provenance models for workflows. The database community has also studied an intriguing new question, i.e., why is $t \notin Q(D)$? This *missing answer* problem is also known as *why-not provenance* [65] and is an area of active research [66–72]. We will return to this question briefly in Sect. 3.4.

The comprehensive survey by Cheney et al. [47] classifies data provenance approaches into two broad categories called *lazy* and *eager*, respectively. In the lazy (or *non-annotation*) approaches, provenance is computed only on demand by examining and analyzing the input data D, the answers A, and the query Q. No

[8]In abstract algebra, a *semiring* is a structure $(K, +, \cdot, 0, 1)$ with binary operations "+" (addition) and "·" (multiplication) over an underlying set K satisfying, for all $x, y, z \in K$, these axioms: $x+y=y+x$; $x+0=0+x=x$; $x \cdot 1 = 1 \cdot x = x$; $x \cdot 0 = 0 \cdot x = 0$; $x \cdot (y+z) = x \cdot y + x \cdot z$; $(x+y) \cdot z = x \cdot z + y \cdot z$. If $x \cdot y = y \cdot x$, the semiring is *commutative*. Instead of $x \cdot y$ we can write xy.

changes are made to any of these. In contrast, the eager (or *annotation-based*) approaches use an annotated input database D' which is then evaluated using a rewritten "provenance-enabled" query Q' in order to obtain an answer table A' with provenance annotations. In the remainder of the paper, we focus on eager provenance approaches. In Appendix we illustrate the exact nature of Q' and the provenance-annotated query answers A' via prototypical implementations of the running example discussed next.

Mixed forms that combine aspects of eager and lazy approaches also exist, e.g., [60]. Several systems such as Perm [73], GProM [74], Ariadne [75], and PROV-Trace [76] compute provenance on demand through provenance-enabled replay of operations. These systems therefore do not fit neatly into the two categories proposed in [47]: On one hand, they appear lazy since provenance is not captured when evaluating a query but only later, if and when provenance is explicitly requested. On the other hand, the technique used for computing provenance is based on provenance-enabled queries that propagate annotations, i.e., the eager approach. The GProM system stands out since it is the first to support provenance tracking for *updates* (and transactions) based on MV-semirings, an extension of the semiring model with embedded multiversion history [77].

3.2 Running Example: The Three-Hop Query (thop)

Consider a database table hop that stores possible links between nodes in a network [78]. We might want to know which pairs of nodes are reachable with precisely three hops. Figure 9 shows this thop (Three-Hop) query in alternative but equivalent

(a)

$$\mathsf{thop}(S,T) \;:\!-\; \mathsf{hop}(S,U), \mathsf{hop}(U,V), \mathsf{hop}(V,T).$$

$$\begin{aligned}\mathsf{thop} := \pi_{S,T}(\quad & \rho_{T\to U}(\mathsf{hop}) \\ & \bowtie \rho_{S\to U, T\to V}(\mathsf{hop}) \\ & \bowtie \rho_{S\to V}(\mathsf{hop})\ )\end{aligned}$$

```
SELECT h1.S, h3.T
FROM   hop h1, hop h2, hop h3
WHERE  h1.T = h2.S
AND    h2.T = h3.S ;
```

(b)

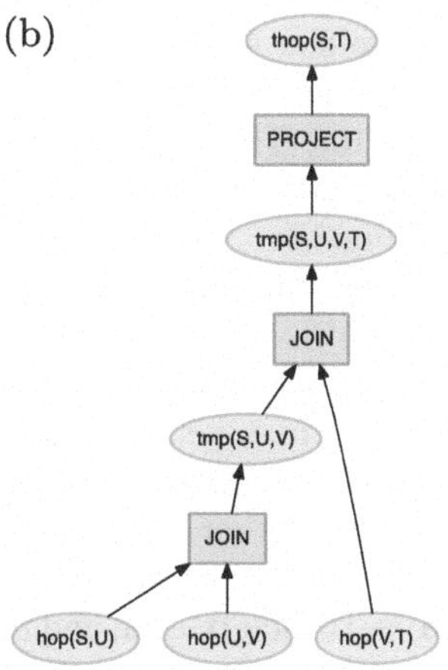

Fig. 9 Three-Hop (thop) query [78], expressed in (**a**) Datalog (*top*), the Relational Algebra (*middle*), and SQL (*bottom*). (**b**) This query can also be considered a "mini-workflow" combining three copies of the hop relation via joins (denoted $\bowtie$ in the algebra), followed by a projection (denoted π) to yield the output relation thop

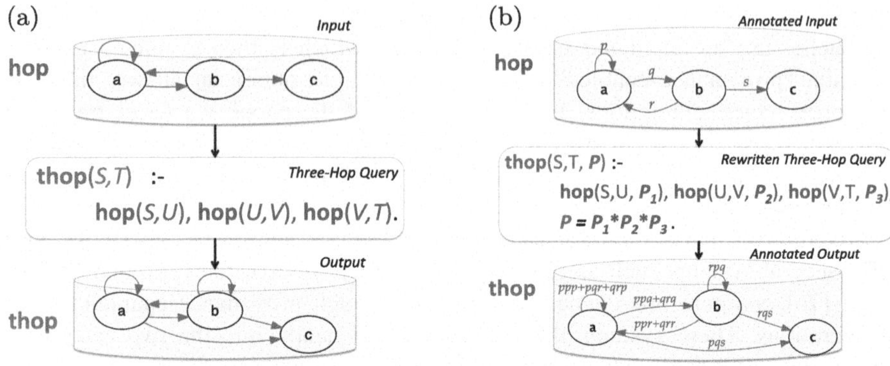

Fig. 10 Three-Hop Example [78]: (**a**) The hop relation (*blue*) stores possible links in a network. The thop query (*center*) returns an output relation (*red*) consisting of all pairs (S, T) that can be connected by three hops $S \overset{\mathsf{hop}}{\rightarrow} U \overset{\mathsf{hop}}{\rightarrow} V \overset{\mathsf{hop}}{\rightarrow} T$ (*bottom*). Typical provenance queries are: *Why* is thop(a,b) in the output, and *how* has thop(a,b) been derived from the input? (**b**) The provenance-annotated input is processed via a rewritten thop query, returning a provenance-annotated output that answers those questions: The provenance polynomial $p^2q + q^2r$ that annotates the thop(a,b) edge (*bottom*) means that there are two distinct ways from a to b using three hops: by using the p hop twice and the q hop once (p^2q), or alternatively, by q, r, and q again (q^2r) (Color figure online)

notations: as a Datalog query, a relational algebra query, and a SQL query. Finally, Fig. 9b shows the same query in the form of a (relational algebra) operator tree. Using operator trees allows us to view a database query Q as a kind of workflow W_Q (or prospective provenance), and apply notions and techniques from Sect. 2. As mentioned before, the processing steps (actors) in workflows are usually considered black boxes. In contrast, in database queries, the semantics of query operators is completely known and available for analysis and query rewriting, making them white box actors that support fine-grained provenance capture. Now consider the thop query from Fig. 9 applied to a concrete input database D as depicted in Fig. 10a. The input relation hop is shown as a directed graph (with blue edges). From this, the query computes a new graph (with red edges), shown at the bottom of Fig. 10a. Note that the hop input graph has no direct link from a to c, while the thop result graph has such as link. Typical provenance queries are:

Why is some tuple t in the output relation thop, and *how* has it been derived from the input relation hop? Consider the result tuple $t = (\mathsf{a}, \mathsf{b})$ in thop. What is the *lineage* of t, i.e., what are the hop tuples that contributed to the derivation of the result $\mathsf{thop}(\mathsf{a}, \mathsf{b})$? Looking at the hop graph, we see that one can go from a to b using different edges from the input hop table, e.g., use the self-loop $\mathsf{a} \rightarrow \mathsf{a}$ twice, followed by the hop $\mathsf{a} \rightarrow \mathsf{b}$, for a total of three hops. Another solution is to use $\mathsf{a} \rightarrow \mathsf{b}$, then $\mathsf{b} \rightarrow \mathsf{a}$, and finally $\mathsf{a} \rightarrow \mathsf{b}$ one more time.

Figure 10b shows the same input database D_{hop} with a small but important modification: the edges in the hop relation are *annotated* with unique identifiers from an underlying set (or namespace) $X = \{p, q, r, s\}$. Thus, we can explain why $\mathsf{thop}(\mathsf{a}, \mathsf{b})$ is in the answer simply by referring to the named edges: p, p, q is a three-hop from a to b, and q, r, q is another three-hop, i.e., $\mathsf{a} \xrightarrow{p} \mathsf{a} \xrightarrow{p} \mathsf{a} \xrightarrow{q} \mathsf{b}$ is the first solution, and $\mathsf{a} \xrightarrow{q} \mathsf{b} \xrightarrow{r} \mathsf{a} \xrightarrow{q} \mathsf{b}$ is the only other solution. A shorthand for the provenance-annotated result is thus "thop(a, b) : $p^2q + q^2r$". The *provenance polynomial* $p^2q + q^2r$ states why and how the query answer thop(a,b) was obtained from the hop input table. The addition "+" in the provenance polynomial corresponds to a logical disjunction ($\vee$) since there are two solutions to go from a to b using exactly three hop edges. Each solution consists of a product "$\cdot$" of input tuples, corresponding to a logical conjunction ($\wedge$), i.e., $p \cdot p \cdot q$ and $q \cdot r \cdot q$. In the underlying provenance semiring [63], the product and sum operations are commutative, hence the shorter polynomial representation $p^2q + q^2r$ can be used.

3.3 *The Great Unification: Provenance Semirings*

The representation of database provenance using abstract polynomials over annotated input databases was developed by Green et al. in [63]; an introduction and overview is given in [78]. It is beyond the scope of this paper to elaborate on the details of that framework and its theoretical results (e.g., the "Fundamental Theorem"). However, using the running example, we can get a first idea of the elegance and power of the semiring approach. Figure 11a depicts the thop answer table with its six output tuples (corresponding to the six red thop edges in Fig. 10). Each of the tuples in the provenance-annotated answer A' carries a provenance annotation which is obtained by executing a rewritten query Q' on an annotated input database D' (see also Appendix). The most fine-grained provenance annotations are shown in the right-most column containing polynomials over the provenance semiring $\mathbb{N}[X]$. The other columns correspond to coarser provenance abstractions: e.g., $\mathbb{B}[X]$ is the semiring of Boolean provenance polynomials, Trio(X) is the provenance semiring used in the Trio system [79], while Why(X) and Lin(X) correspond to the *why-provenance* and *lineage* model in [60, 62], respectively.

(a)

thop	S	T	Lin(X)	Why(X)	Trio(X)	$\mathbb{B}[X]$	$\mathbb{N}[X]$
	a	a	pqr	$p + pqr$	$p + 2pqr$	$p^3 + pqr$	$p^3 + 2pqr$
	a	b	pqr	$pq + qr$	$pq + qr$	$p^2q + q^2r$	$p^2q + q^2r$
	a	c	pqs	pqs	pqs	pqs	pqs
	b	a	pqr	$pr + qr$	$pr + qr$	$p^2r + qr^2$	$p^2r + qr^2$
	b	b	pqr	pqr	pqr	pqr	pqr
	b	c	qrs	qrs	qrs	qrs	qrs

(b)

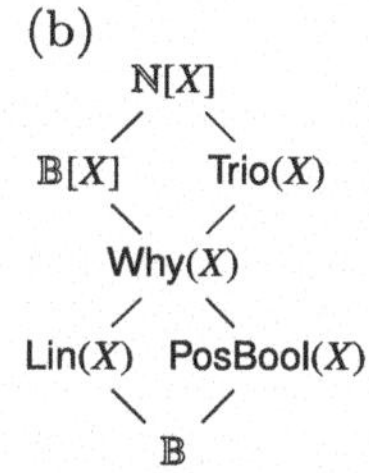

Fig. 11 Three-Hop Example (cont'd): (**a**) Provenance-annotated thop answer with five kinds of provenance. (**b**) The hierarchy among provenance models [78]: the finest-grain model $\mathbb{N}[X]$ subsumes other models such as Trio(X), Why(X), and Lin(X) *below*. For example, the Lin(X) model for thop(a, b) only states that the hop edges p, q, r are in the lineage, while the $\mathbb{N}[X]$ model states exactly *how* those edges need to be combined

The lattice in Fig. 11 shows the degree of "informedness" of the different provenance models (i.e., how "fine-grained" they are, relative to one another): as one moves down the lattice, provenance information becomes coarser. In our example, the $\mathbb{N}[X]$ provenance of thop(a, b) is $p^2q + q^2r$ telling us (1) that there are exactly two ways to obtain the answer, and (2) what those two ways are (one way uses the *p* edge twice and *q* once; the other uses *q* twice and *r* once). When looking instead at $\mathbb{B}[X]$, the coefficients are dropped from the polynomial, e.g., the provenance of thop(a, a) is $p^3 + 2pqr$ in $\mathbb{N}[X]$, but becomes $p^3 + pqr$ in $\mathbb{B}[X]$. Similarly, in Trio(X), exponents are dropped, in Why(X) coefficients *and* exponents are dropped, and in Lin(X) only the (flat) union of tuples *pqr* remains to describe the lineage of thop(a, a), i.e., these three edges were used in the derivation, but it is not stated *how* they need to be put together to derive a three-hop from a to a.

The "Fundamental Theorem" [78] intuitively states that for positive relational algebra queries one can swap the order of query evaluation and application of a semiring homomorphism. For example, consider an input database with annotations $p, q, r, \ldots$ that represent Boolean variables that can be either *true* or *false*, indicating whether the so-annotated tuple is or isn't true in the modeled world. In order to explore the answers to a query Q in different possible worlds (i.e., under different truth assignments to the Booleans), we could run the query Q once for each such possible world. Alternatively, we can execute the provenance-enabled query Q' once (and for all) to obtain provenance polynomials in $\mathbb{N}[X]$ as depicted in Fig. 11a. To obtain the different possible worlds, we then just reinterpret the provenance-polynomials as Boolean expressions ("·" as "$\wedge$" and "+" as "$\vee$") and simplify those Boolean expressions. Both routes (Boolean assignment followed by query evaluation or vice versa) will yield the same result.

Appendix contains another example, where the input annotations represent tuple cardinalities in the relational model with multiset (bag) semantics. We can evaluate the query under the bag semantics to obtain the result cardinalities (Fig. 15c, d). Alternatively, we can "plug in" the input cardinalities into the abstract provenance polynomials in Fig. 15b and then evaluate those polynomials to arrive at the same numbers as in Fig. 15d.

3.4 Unifying Why and Why-Not Provenance Through Games

The elegant and powerful provenance semiring approach by Green et al. [63, 78] subsumes and situates many earlier database provenance models. However, one shortcoming of those approaches is that they are limited to *positive* queries only, i.e., they cannot handle queries with negation. On the other hand, if a provenance approach can be devised that can answer queries with negation, then such an approach would also solve the missing answers or why-not provenance problem: Asking why is thop(c, a) *not* in the answer is then equivalent to asking: why is $\neg$thop(c, a) true over the given database. Figure 12a depicts a solved *provenance game* for thop(a,a). This approach was developed by Köhler et al. [70] and contains the provenance semiring approach as a special case, see Fig. 12b. The key idea is to

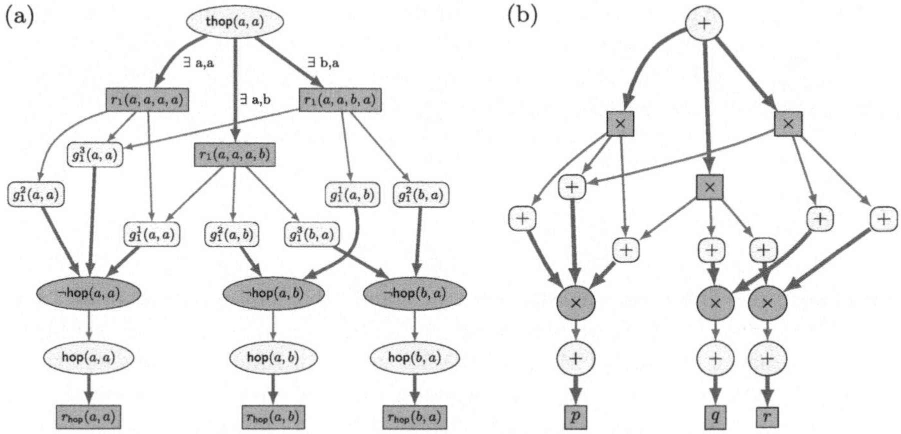

Fig. 12 Why (and how) is thop(a,a) in the query answer? (**a**) The solved provenance game [70] shows that one can find three different instances of the thop Datalog rule, such that those rule instances are satisfied. (**b**) The solved game DAG can be abstracted and expanded into a tree to yield the provenance polynomial for thop(a,a): $p^3 + 2pqr$

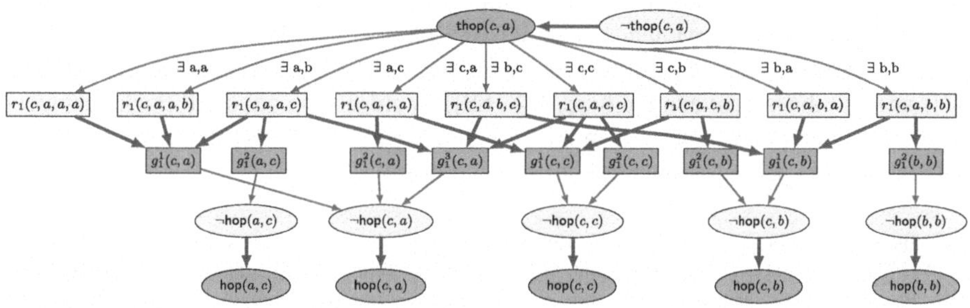

Fig. 13 Why-not provenance for thop(c,a) using provenance games [70]. The graph enumerates all (failed) attempts to prove thop(c,a) using the thop query over the given hop database. This structure can also be used to propose changes to the database such that thopc(c,a) will be in the answer

view query evaluation $A = Q(D)$ as a game between two players who argue whether or not tuple $t \in A$.[9] The game can be defined in such a way that whoever is right about the claim can force a win [70]. Then the provenance (or *justifications*) for a claim about $t \in A$ can be obtained from a solved game graph such as the one in Fig. 12a.

A key advantage of this approach is that it treats why and why-not provenance uniformly: Fig. 13 depicts a solved query evaluation game establishing why

[9]Query evaluation games [80] have been considered before, e.g., by Hintikka [81]. However, the idea of using games for provenance was inspired more recently by revisiting the game normal form [82] for well-founded Datalog.

thop(c,a) is *not* in the answer. The solved game graph contains the equivalent of all failed proof attempts using the rules of the query (corresponding to failed SLD(NF) trees [83]) and can be used, e.g., to determine how the given database can be "fixed" so that thop(c,a) becomes true after all.

4 Conclusions

Provenance is a flourishing research area in many subdisciplines of computer science. The scientific workflow community has contributed to the development of the Open Provenance Model (OPM) and its W3C successor PROV [84]. As described in this chapter, two main forms of provenance can be distinguished in workflows, i.e., prospective and retrospective provenance. When combined in a single model of provenance (possibly enriched with temporal information), powerful provenance queries can be answered. The database community has developed another set of provenance models which abstract tuple derivations through relational queries (or Datalog rules). The provenance semiring model introduced by Green et al. [63] elegantly subsumes many earlier provenance models for positive queries. Why-not (or missing answer) provenance is an active area of research.

In this brief tour, many interesting topics in workflow provenance (e.g., [22, 85–87]) and database provenance (e.g., [88, 89]) could not be covered. For overviews and surveys on provenance and workflow see, e.g., [13–15, 84]. For provenance in databases, [47] provides an excellent starting point.

Acknowledgements This work was supported in part by NSF grants ACI-1430508, DBI-{1147273, 1356751}, IIS-1118088, and SMA-1439603. With special thanks to Shawn Bowers, Timothy McPhillips, Manish K. Anand, Víctor Cuevas-Vicenttín, Saumen Dey, Lei Dou, Sven Köhler, Sean Riddle, and Daniel Zinn for fruitful years of collaboration on scientific workflows and database provenance. Also special thanks to Boris Glavic for comments on an earlier draft of this paper and for his collaboration on and implementation of games for why-not provenance.

Appendix: Query Rewriting for Provenance Annotations

The key ideas behind the query rewriting in the semiring annotation approach [63] can be nicely illustrated using some simple prototypical implementations.[10] Figure 14 depicts two variants of the three-hop query [78] used earlier in the paper. The first variant (Fig. 14a) uses unique tuple-ids and a symbolic representation of the product operation in the $\mathbb{N}[X]$ semiring. Lists of such products are used to represent the sum of products form in Fig. 14b. In Fig. 14c the same query is used, but now hop represents a multiset (bag semantics), so tuples are annotated with cardinalities

[10]The example code is available from github.com/idaks/tour-de-provenance.

A Query Rewriting for Provenance Annotations

(a)

```
% hop(X,Y) relation with unique tuple-ids p,q,r,s:
hop(a,a, p).
hop(a,b, q).
hop(b,a, r).
hop(b,c, s).

% Rewritten Three-Hop Query:
thop_aux(X,Y, P) :-
        hop(X,U, P1), hop(U,V, P2), hop(V,Y, P3),
        P = P1*P2*P3.

% For each X,Y pair, aggregate all provenance annotations.
thop(X,Y, Ps) :-
        bagof( P, thop_aux(X,Y, P), Ps ).

% Backtracking loop to generate all thop(X,Y) with provenance:
:-      thop(X,Y, Ps), format("thop(~w,~w) : ~w~n", [X,Y,Ps]), fail
    ;   true.
```

(b)

```
Welcome to SWI-Prolog (Multi-threaded, 64 bits, Version 6.0.2)
Copyright (c) 1990-2011 University of Amsterdam, VU Amsterdam
SWI-Prolog comes with ABSOLUTELY NO WARRANTY. This is free software,
and you are welcome to redistribute it under certain conditions.
Please visit http://www.swi-prolog.org for details.

For help, use ?- help(Topic). or ?- apropos(Word).

?- [thop].
thop(a,a) : [p*p*p,p*q*r,q*r*p]
thop(a,b) : [p*p*q,q*r*q]
thop(a,c) : [p*q*s]
thop(b,a) : [r*p*p,r*q*r]
thop(b,b) : [r*p*q]
thop(b,c) : [r*q*s]
% thop compiled 0.00 sec, 8 clauses
true.

?-
```

(c)

```
:- use_module(library(lists)). % for arithmetic list sum

% hop(X,Y) multiset with cardinalities:
hop(a,a, 1).
hop(a,b, 4).
hop(b,a, 2).
hop(b,c, 3).

% Rewritten Three-Hop Query:
thop_aux(X,Y,  N) :-
        hop(X,U, N1), hop(U,V, N2), hop(V,Y, N3),
        N is N1*N2*N3.

% For each X,Y pair, aggregate all provenance annotations:
thop(X,Y, Ns) :-
        bagof( N , thop_aux(X,Y,N), Ns ).

% Backtracking loop to generate all thop(X,Y) with provenance:
:-      thop(X,Y, Ns), sumlist(Ns,S), % arithmetic list sum
        format('thop(~w,~w) : ~w~n', [X,Y,S]), fail
    ;   true.
```

(d)

```
Welcome to SWI-Prolog (Multi-threaded, 64 bits, Version 6.0.2)
Copyright (c) 1990-2011 University of Amsterdam, VU Amsterdam
SWI-Prolog comes with ABSOLUTELY NO WARRANTY. This is free software,
and you are welcome to redistribute it under certain conditions.
Please visit http://www.swi-prolog.org for details.

For help, use ?- help(Topic). or ?- apropos(Word).

?- [thop2].
%   library(error) compiled into error 0.00 sec, 79 clauses
%  library(lists) compiled into lists 0.00 sec, 179 clauses
thop(a,a) : 17
thop(a,b) : 36
thop(a,c) : 12
thop(b,a) : 18
thop(b,b) : 8
thop(b,c) : 24
% thop2 compiled 0.01 sec, 190 clauses
true.

?-
```

Fig. 14 Three-Hop Example [78] prototypically implemented in SWI-Prolog: (**a**) The input relation hop is annotated with unique tuple-ids. The rewritten view thop_aux adds the symbolic product $\mathsf{P} = \mathsf{P}_1 \cdot \mathsf{P}_2 \cdot \mathsf{P}_3$, combining the provenance annotations P_i of all hop tuples being joined. Aggregation with bagof/3 (instead of setof/3) is used to collect all provenance. (**b**) Running the code from (**a**) generates the provenance polynomials. (**c**) Similar to (**a**) but now hop is a multiset with cardinality annotations. The provenance of the thop result is calculated as the sum of the *arithmetic* product of the input cardinalities. (**d**) Running the code from (**c**) generates the result cardinalities

(how many times a tuple is in the multiset). The resulting cardinalities in the thop result relation are obtained by computing the sum of the arithmetic products of the cardinalities of hop tuples being joined to obtain the annotated thop tuples. The use of bag semantics (via the built-in aggregation predicate bagof/3, rather than setof/3) is essential to obtain the correct cardinalities.

In Fig. 15 the same thop query with provenance is implemented in SQLite, again first using provenance polynomials over the $\mathbb{N}[X]$ semiring (using symbolic tuple-ids, represented as strings). The second variant in Fig. 15c, d uses multiset semantics where tuple cardinalities are represented numerically in an additional column. The result cardinalities are then obtained via a summation over the (arithmetic) products of thop annotations.

(a)

```
CREATE TABLE hop (S text, T text, P text);

INSERT INTO hop VALUES ("a","a", "p");
INSERT INTO hop VALUES ("a","b", "q");
INSERT INTO hop VALUES ("b","a", "r");
INSERT INTO hop VALUES ("b","c", "s");

CREATE VIEW thop AS
    SELECT h1.S, h3.T, h1.P||'*'||h2.P||'*'||h3.P AS P
    FROM   hop h1, hop h2, hop h3
    WHERE  h1.T = h2.S AND h2.T = h3.S ;

SELECT   S, T, group_concat(P, ' + ')
FROM     thop
GROUP BY S, T;
```

(b)

```
$ sqlite3 -init thop.sql
-- Loading resources from thop.sql
S           T           group_concat(P, ' + ')
----------  ----------  ----------------------
a           a           p*p*p + p*q*r + q*r*p
a           b           p*p*q + q*r*q
a           c           p*q*s
b           a           r*p*p + r*q*r
b           b           r*p*q
b           c           r*q*s

SQLite version 3.8.11.1 2015-07-29 20:00:57
Enter ".help" for usage hints.
sqlite>
```

(c)

```
CREATE TABLE hop (S text, T text, P number);

INSERT INTO hop VALUES ("a","a", 1);
INSERT INTO hop VALUES ("a","b", 4);
INSERT INTO hop VALUES ("b","a", 2);
INSERT INTO hop VALUES ("b","c", 3);

CREATE VIEW thop AS
    SELECT h1.S, h3.T, h1.P * h2.P * h3.P AS P
    FROM   hop h1, hop h2, hop h3
    WHERE  h1.T = h2.S AND h2.T = h3.S ;

SELECT   S, T, sum(P)
FROM     thop
GROUP BY S, T;
```

(d)

```
$ sqlite3 -init thop2.sql
-- Loading resources from thop2.sql
S           T           sum(P)
----------  ----------  ----------
a           a           17
a           b           36
a           c           12
b           a           18
b           b           8
b           c           24

SQLite version 3.8.11.1 2015-07-29 20:00:57
Enter ".help" for usage hints.
sqlite>
```

Fig. 15 Three-Hop Example [78] prototypically implemented in SQLite via query rewritings: (**a**) the rewritten view `thop` adds a column that symbolically"multiplies" the provenance of the `hop` tuples being joined; (**b**) running the aggregation query from (**a**) yields the provenance polynomials from $\mathbb{N}[X]$; (**c**) variant similar to (**a**) but for bag semantics; (**d**) running the aggregation from (**c**) yields the expected multiplicities

References

1. Wedel, M.J.: A monument of inefficiency: the presumed course of the recurrent laryngeal nerve in sauropod dinosaurs. Acta Palaeontol. Pol. **57**(2), 251–256 (2011)
2. Dobzhansky, T.: Nothing in biology makes sense except in the light of evolution. Am. Biol. Teach. **35**(3), 125–129 (1973)
3. Hey, T., Tansley, S., Tolle, K. (eds.): The fourth paradigm: data-intensive scientific discovery. Microsoft Research Redmond, WA (2009)
4. GCIS: Global Change Information System (2015). http://data.globalchange.gov/
5. Melillo, J.M., Richmond, T.T., Yohe, G.W. (eds.): Climate Change Impacts in the United States: The Third National Climate Assessment. U.S. Global Change Research Program (2014). doi:10.7930/J0Z31WJ2
6. Tilmes, C., Fox, P., Ma, X.L., McGuinness, D.L., Privette, A.P., Smith, A., Waple, A., Zednik, S., Zheng, J.G.: Provenance representation for the national climate assessment in the global change information system. IEEE Trans. Geosci. Remote Sens. **51**(11), 5160–5168 (2013)
7. Sadiq, S.: Handbook of Data Quality. Springer, Berlin (2013)
8. Mann, M.E., Zhang, Z., Hughes, M.K., Bradley, R.S., Miller, S.K., Rutherford, S., Ni, F.: Proxy-based reconstructions of hemispheric and global surface temperature variations over the past two millennia. Proc. Natl. Acad. Sci. **105**(36), 13252–13257 (2008)

9. Hills, D.J., Downs, R.R., Duerr, R., Goldstein, J.C., Parsons, M.A., Ramapriyan, H.K.: The importance of data set provenance for science. Eos **96** (2015). doi:10.1029/2015EO040557
10. Eisenman, I., Meier, W.N., Norris, J.R.: A spurious jump in the satellite record: has Antarctic sea ice expansion been overestimated? Cryosphere **8**(4), 1289–1296 (2014)
11. Stevens, L.: Texas Summer 2011: Record Heat and Drought (2013). GCIS metadata record with provenance. Accessed 12 Dec 2015
12. Ludäscher, B., Bowers, S., McPhillips, T.: Scientific workflows. In: Özsu, T., Liu, L. (eds.) Encyclopedia of Database Systems. Springer, Berlin (2009)
13. Cuevas-Vicenttín, V., Dey, S., Köhler, S., Riddle, S., Ludäscher, B.: Scientific workflows and provenance: introduction and research opportunities. Datenbank-Spektrum **12**(3), 193–203 (2012)
14. Davidson, S.B., Boulakia, S.C., Eyal, A., Ludäscher, B., McPhillips, T.M., Bowers, S., Anand, M.K., Freire, J.: Provenance in scientific workflow systems. IEEE Data Eng. Bull. **30**(4), 44–50 (2007)
15. Bowers, S.: Scientific workflow, provenance, and data modeling challenges and approaches. J. Data Semant. **1**(1), 19–30 (2012)
16. Ludäscher, B., Altintas, I., Bowers, S., Cummings, J., Critchlow, T., Deelman, E., Roure, D.D., Freire, J., Goble, C., Jones, M., Klasky, S., McPhillips, T., Podhorszki, N., Silva, C., Taylor, I., Vouk, M.: Scientific process automation and workflow management. In: Shoshani, A., Rotem, D. (eds.) Scientific Data Management. Chapman & Hall/CRC, London/Boca Raton (2009)
17. McPhillips, T., Bowers, S., Zinn, D., Ludäscher, B.: Scientific workflow design for mere mortals. Futur. Gener. Comput. Syst. **25**(5), 541–551 (2009)
18. Dou, L., Cao, G., Morris, P.J., Morris, R.A., Ludäscher, B., Macklin, J.A., Hanken, J.: Kurator: a kepler package for data curation workflows. Proc. Comput. Sci. **9**, 1614–1619 (2012). Demo video at http://youtu.be/DEkPbvLsud0
19. Ludäscher, B., Altintas, I., Berkley, C., Higgins, D., Jaeger, E., Jones, M., Lee, E.A., Tao, J., Zhao, Y.: Scientific workflow management and the Kepler system. Concurr. Comput. Pract. Experience **18**(10), 1039–1065 (2006)
20. Bowers, S., McPhillips, T., Riddle, S., Anand, M.K., Ludäscher, B.: Kepler/pPOD: scientific workflow and provenance support for assembling the tree of life. In: Provenance and Annotation of Data and Processes (IPAW), pp. 70–77. Springer, Berlin, Heidelberg (2008)
21. Anand, M.K., Bowers, S., Ludäscher, B.: Provenance browser: displaying and querying scientific workflow provenance graphs. In: IEEE International Conference on Data Engineering (ICDE), pp. 1201–1204 (2010)
22. Zinn, D., Ludäscher, B.: Abstract provenance graphs: anticipating and exploiting schema-level data provenance. In: Provenance and Annotation of Data and Processes, pp. 206–215. Springer, Berlin, Heidelberg (2010)
23. Moreau, L., Ludäscher, B., Altintas, I., Barga, R.S., Bowers, S., Callahan, S., Chin, G., Clifford, B., Cohen, S., Cohen-Boulakia, S., Davidson, S., Deelman, E., Digiampietri, L., Foster, I., Freire, J., Frew, J., Futrelle, J., Gibson, T., Gil, Y., Goble, C., Golbeck, J., Groth, P., Holland, D.A., Jiang, S., Kim, J., Koop, D., Krenek, A., McPhillips, T., Mehta, G., Miles, S., Metzger, D., Munroe, S., Myers, J., Plale, B., Podhorszki, N., Ratnakar, V., Santos, E., Scheidegger, C., Schuchardt, K., Seltzer, M., Simmhan, Y.L., Silva, C., Slaughter, P., Stephan, E., Stevens, R., Turi, D., Vo, H., Wilde, M., Zhao, J., Zhao, Y.: Special issue: the first provenance challenge. Concurr. Comput. Pract. Experience **20**(5), 409–418 (2008)
24. Moreau, L., Freire, J., Futrelle, J., McGrath, R.E., Myers, J., Paulson, P.: The open provenance model: an overview. In: Provenance and Annotation of Data and Processes, pp. 323–326. Springer, Berlin (2008)
25. Moreau, L., Clifford, B., Freire, J., Futrelle, J., Gil, Y., Groth, P., Kwasnikowska, N., Miles, S., Missier, P., Myers, J., Plale, B., Simmhan, Y., Stephan, E., den Bussche, J.V.: The open provenance model core specification (v1. 1). Futur. Gener. Comput. Syst. **27**(6), 743–756 (2011)

26. Moreau, L., Missier, P., Belhajjame, K., B'Far, R., Cheney, J., Coppens, S., Cresswell, S., Gil, Y., Groth, P., Klyne, G., Lebo, T., McCusker, J., Miles, S., Myers, J., Sahoo, S., Tilmes, C.: The PROV data model. W3C Technical Report (2012). https://www.w3.org/TR/prov-dm/
27. Heinis, T., Alonso, G.: Efficient lineage tracking for scientific workflows. In: SIGMOD, pp. 1007–1018. ACM, New York (2008)
28. Chapman, A.P., Jagadish, H.V., Ramanan, P.: Efficient provenance storage. In: Proceedings of the 2008 ACM SIGMOD International Conference on Management of Data, pp. 993–1006. ACM, New York (2008)
29. Anand, M.K., Bowers, S., McPhillips, T., Ludäscher, B.: Efficient provenance storage over nested data collections. In: International Conference on Extending Database Technology (EDBT), pp. 958–969. ACM, New York (2009)
30. Anand, M.K., Bowers, S., Ludäscher, B.: A navigation model for exploring scientific workflow provenance graphs. In: 4th Workshop on Workflows in Support of Large-Scale Science (WORKS) (2009)
31. Anand, M.K., Bowers, S., Ludäscher, B.: Techniques for efficiently querying scientific workflow provenance graphs. In: EDBT, vol. 10, pp. 287–298 (2010)
32. Anand, M.K., Bowers, S., Ludäscher, B.: Database support for exploring scientific workflow provenance graphs. In: Scientific and Statistical Database Management, pp. 343–360. Springer, Berlin, Heidelberg (2012)
33. Garijo, D., Gil, Y.: A new approach for publishing workflows: abstractions, standards, and linked data. In: 6th Workshop on Workflows in Support of Large-Scale Science (WORKS) (2011)
34. Missier, P., Dey, S., Belhajjame, K., Cuevas-Vicenttín, V., Ludäscher, B.: D-PROV: extending the prov provenance model with workflow structure. In: 5th USENIX Workshop on the Theory and Practice of Provenance (TaPP) (2013)
35. Dey, S., Köhler, S., Bowers, S., Ludäscher, B.: Datalog as a lingua franca for provenance querying and reasoning. In: Workshop on the Theory and Practice of Provenance (TaPP), Boston, MA (2012)
36. Pham, Q., Malik, T., Glavic, B., Foster, I.: LDV: light-weight database virtualization. In: International Conference on Data Engineering (ICDE), pp. 1179–1190 (2015)
37. Kwasnikowska, N., Moreau, L., Bussche, J.V.D.: A formal account of the open provenance model. ACM Trans. Web (TWEB) **9**(2), 10:1–10:44 (2015)
38. Dey, S., Riddle, S., Ludäscher, B.: Provenance analyzer: exploring provenance semantics with logic rules. In: 5th USENIX Workshop on the Theory and Practice of Provenance (TaPP) (2013)
39. Dijkstra, E.W.: Hamming's exercise in SASL. EWD-792 (1981)
40. Hemmendinger, D.: The "Hamming problem" in prolog. ACM SIGPLAN Not. **23**(4), 81–86 (1988)
41. Köhler, S., Ludäscher, B., Smaragdakis, Y.: Declarative datalog debugging for mere mortals. In: Datalog in Academia and Industry, pp. 111–122. Springer, Berlin, Heidelberg (2012)
42. Koop, D., Santos, E., Bauer, B., Troyer, M., Freire, J., Silva, C.T.: Bridging workflow and data provenance using strong links. In: Gertz, M., Ludäscher, B. (eds.) Scientific and statistical database management (SSDBM). Lecture Notes in Computer Science, vol. 6187, Springer, Berlin (2010)
43. Bowers, S., McPhillips, T., Ludäscher, B.: Declarative rules for inferring fine-grained data provenance from scientific workflow execution traces. In: International Provenance and Annotation Workshop (IPAW), pp. 82–96. Springer (2012)
44. Dey, S., Belhajjame, K., Koop, D., Song, T., Missier, P., Ludäscher, B.: UP & DOWN: improving provenance precision by combining workflow-and trace-level information. In: 6th USENIX Workshop on the Theory and Practice of Provenance (TaPP), Cologne (2014)
45. Dey, S., Belhajjame, K., Koop, D., Raul, M., Ludäscher, B.: Linking prospective and retrospective provenance for scripts. In: 7th USENIX Workshop on the Theory and Practice of Provenance (TaPP), Edinburgh (2015)

46. McPhillips, T., Bowers, S., Belhajjame, K., Ludäscher, B.: Retrospective provenance without a runtime provenance recorder. In: 7th USENIX Workshop on the Theory and Practice of Provenance (TaPP), Edinburg (2015)
47. Cheney, J., Chiticariu, L., Tan, W.: Provenance in databases: why, how, and where. Found. Trends Databases **1**(4), 379–474 (2009)
48. Cohen, S., Cohen-Boulakia, S., Davidson, S.: Towards a model of provenance and user views in scientific workflows. In: Data Integration in the Life Sciences (DILS), pp. 264–279. Springer, Berlin
49. Tan, W.C.: Provenance in databases: past, current, and future. IEEE Data Eng. Bull. **30**(4), 3–12 (2007)
50. Bowers, S., Ludäscher, B.: Actor-oriented design of scientific workflows. In: Conceptual Modeling (ER). Lecture Notes in Computer Science, vol. 3716, pp. 369–384. Springer, Berlin (2005)
51. Biton, O., Cohen-Boulakia, S., Davidson, S.B., Hara, C.S.: Querying and managing provenance through user views in scientific workflows. In: International Conference on Data Engineering (ICDE), pp. 1072–1081. IEEE, New York (2008)
52. Murta, L., Braganholo, V., Chirigati, F., Koop, D., Freire, J.: noWorkflow: Capturing and analyzing provenance of scripts. In: Provenance and Annotation of Data and Processes (IPAW), pp. 71–83. Springer, Berlin (2014)
53. Buneman, P., Tan, W.C.: Provenance in databases (Tutorial Outline). In: SIGMOD, pp. 1171–1173. ACM, New York (2007)
54. Amsterdamer, Y., Davidson, S.B., Deutch, D., Milo, T., Stoyanovich, J., Tannen, V.: Putting lipstick on pig: enabling database-style workflow provenance. Proc. VLDB Endow. **5**(4), 346–357 (2011)
55. Abiteboul, S., Hull, R., Vianu, V.: Foundations of Databases. Addison-Wesley, Reading, MA (1995)
56. Deutsch, A., Tannen, V.: Reformulation of XML Queries and Constraints. In: International Conference on Database Theory (ICDT), pp. 225–241. Springer, Berlin (2003)
57. Boncz, P., Grust, T., Van Keulen, M., Manegold, S., Rittinger, J., Teubner, J.: MonetDB/XQuery: a fast XQuery processor powered by a relational engine. In: SIGMOD, pp. 479–490. ACM, New York (2006)
58. Wang, Y.R., Madnick, S.E., et al.: A polygen model for heterogeneous database systems: the source tagging perspective. In: VLDB, vol. 90, pp. 519–538 (1990)
59. Woodruff, A., Stonebraker, M.: Supporting fine-grained data lineage in a database visualization environment. In: International Conference on Data Engineering (ICDE), pp. 91–102. IEEE, New York (1997)
60. Cui, Y., Widom, J., Wiener, J.: Tracing the lineage of view data in a warehousing environment. ACM Trans. Database Systems **25**(2), 179–227 (2000)
61. Chaudhuri, S., Dayal, U.: Data warehousing and OLAP for decision support. ACM Sigmod Rec. **26**(2), 507–508 (1997)
62. Buneman, P., Khanna, S., Tan, W.C.: Why and where: a characterization of data provenance. In: ICDT, pp. 316–330. Springer, Berlin (2001)
63. Green, T., Karvounarakis, G., Tannen, V.: Provenance semirings. In: PODS, pp. 31–40 (2007)
64. Green, T.J., Karvounarakis, G., Tannen, Z.G.I.V.: Provenance in ORCHESTRA. In: Bulletin of the Technical Committee on Data Engineering, vol. 33(3), pp. 9–16. IEEE Computer Society, New York (2010)
65. Chapman, A., Jagadish, H.: Why not? In: SIGMOD, pp. 523–534. ACM, New York (2009)
66. Herschel, M., Hernández, M.A.: Explaining missing answers to SPJUA queries. Proc. VLDB Endow. **3**(1–2), 185–196 (2010)
67. Tran, Q.T., Chan, C.Y.: How to ConQueR Why-Not Questions. In: SIGMOD, ACM, New York (2010), pp. 15–26
68. Geerts, F., Poggi, A.: On database query languages for k-relations. J. Appl. Log. **8**(2), 173–185 (2010)

69. Amsterdamer, Y., Deutch, D., Tannen, V.: On the limitations of provenance for queries with difference. In: TaPP (2011)
70. Köhler, S., Ludäscher, B., Zinn, D.: First-order provenance games. In: In Search of Elegance in the Theory and Practice of Computation. Essays Dedicated to Peter Buneman. Lecture Notes in Computer Science, vol. 8000, pp. 382–399. Springer, Berlin (2013)
71. Bidoit, N., Herschel, M., Tzompanaki, K.: EFQ: why-not answer polynomials in action. Proc. VLDB Endow. **8**(12), 1980–1983 (2015)
72. ten Cate, B., Civili, C., Sherkhonov, E., Tan, W.C.: High-level why-not explanations using ontologies. In: ACM Symposium on Principles of Database Systems (PODS), pp. 31–43. ACM, New York (2015)
73. Glavic, B., Miller, R.J., Alonso, G.: Using SQL for efficient generation and querying of provenance information. In: In Search of Elegance in the Theory and Practice of Computation. Essays Dedicated to Peter Buneman. Lecture Notes in Computer Science, vol. 8000, pp. 291–320. Springer, Berlin (2013)
74. Arab, B., Gawlick, D., Radhakrishnan, V., Guo, H., Glavic, B.: A Generic Provenance Middleware for Queries, Updates, and Transactions. In: 6th USENIX Workshop on the Theory and Practice of Provenance (TaPP) (2014)
75. Glavic, B., Esmaili, K.S., Fischer, P.M., Tatbul, N.: Efficient stream provenance via operator instrumentation. ACM Trans. Internet Tech. **14**(1), 7 (2014)
76. Stamatogiannakis, M., Groth, P., Bos, H.: Decoupling provenance capture and analysis from execution. In: 7th USENIX Workshop on the Theory and Practice of Provenance (TaPP) (2015)
77. Arab, B., Gawlick, D., Krishnaswamy, V., Radhakrishnan, V., Glavic, B.: Formal foundations of reenactment and transaction provenance. Technical Report IIT/CS-DB-2016-01. Illinois Institute of Technology (2016)
78. Karvounarakis, G., Green, T.J.: Semiring-annotated data: queries and provenance. ACM SIGMOD Rec. **41**(3), 5–14 (2012)
79. Benjelloun, O., Sarma, A.D., Halevy, A., Widom, J.: ULDBs: Databases with uncertainty and lineage. In: Proceedings of the 32nd International Conference on Very Large Data Bases, VLDB Endowment, pp. 953–964 (2006)
80. Hodges, W.: Logic and Games. In: Zalta, E.N. (ed.) The Stanford Encyclopedia of Philosophy (2013). http://plato.stanford.edu/entries/logic-games/
81. Hintikka, J.: The Principles of Mathematics Revisited. Cambridge University Press, Cambridge (1996)
82. Flum, J., Kubierschky, M., Ludäscher, B.: Total and partial well-founded datalog coincide. In: ICDT, pp. 113–124 (1997)
83. Apt, K.R., Doets, K.: A new definition of SLDNF-resolution. J. Logic Program. **18**(2), 177–190 (1994)
84. Moreau, L.: The foundations for provenance on the web. Found. Trends Web Sci. **2**(2–3), 99–241 (2010)
85. Missier, P., Paton, N.W., Belhajjame, K.: Fine-grained and efficient lineage querying of collection-based workflow provenance. In: EDBT, pp. 299–310 (2010)
86. Missier, P., Ludäscher, B., Bowers, S., Dey, S., Sarkar, A., Shrestha, B., Altintas, I., Anand, M.K., Goble, C.: Linking multiple workflow provenance traces for interoperable collaborative science. In: 5th Workshop on Workflows in Support of Large-Scale Science (WORKS). IEEE, New York (2010)
87. Köhler, S., Riddle, S., Zinn, D., McPhillips, T., Ludäscher, B.: Improving workflow fault tolerance through provenance-based recovery. In: Scientific and Statistical Database Management, pp. 207–224. Springer, Berlin, Heidelberg (2011)
88. Meliou, A., Gatterbauer, W., Moore, K.F., Suciu, D.: The complexity of causality and responsibility for query answers and non-answers. Proc. VLDB Endow. **4**(1), 34–45 (2010)
89. Salimi, B., Bertossi, L.: From causes for database queries to repairs and model-based diagnosis and back. In: 18th International Conference on Database Theory (ICDT), vol. 31, pp. 342–362. Schloss Dagstuhl–Leibniz-Zentrum für Informatik, Wadern (2015)

The Lifecycle of Provenance Metadata and Its Associated Challenges and Opportunities

Paolo Missier

Abstract This chapter outlines some of the challenges and opportunities associated with adopting provenance principles (Cheney et al., Dagstuhl Reports 2(2):84–113, 2012) and standards (Moreau et al., Web Semant. Sci. Serv. Agents World Wide Web, 2015) in a variety of disciplines, including data publication and reuse, and information sciences.

Keywords Provenance analytics • Provenance data modelling • Provenance lifecycle

Using provenance in a broad diversity of application areas and disciplines entails a number of challenges, including specialising the generic provenance and domain-agnostic data model, PROV. This chapter provides a brief overview of these challenges, using the *provenance lifecycle* framework shown in Fig. 1 as a reference.

1 Provenance Definitions and Model

PROV, the Provenance standard, is a family of specifications released in 2013 by the Provenance Working Group, as a contribution to the Semantic Web suite of technologies at the World Wide Web Consortium [36]. PROV aims to define a *generic* data model for provenance that can be extended, in a principled way, to suit many application areas. The PROV-DM document [34] provides an operational definition of provenance for the community to use and build upon:

> Provenance is defined as a record that describes the people, institutions, entities, and activities involved in producing, influencing, or delivering a piece of data or a thing.

P. Missier (✉)
School of Computing Science, Newcastle University, Newcastle upon Tyne, UK
e-mail: Paolo.Missier@ncl.ac.uk

V.L. Lemieux (ed.), *Building Trust in Information*, Springer Proceedings in Business and Economics, DOI 10.1007/978-3-319-40226-0_8

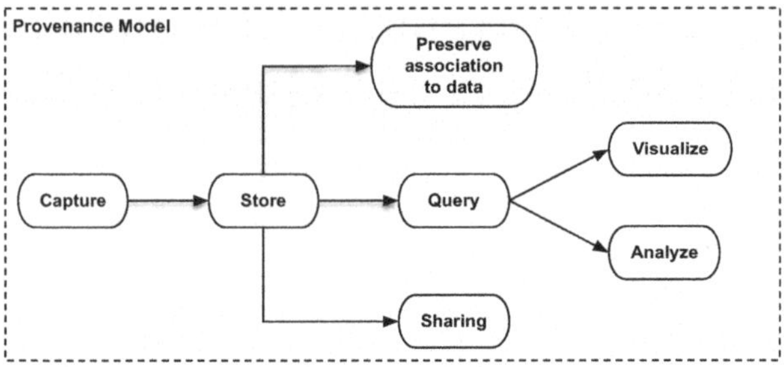

Fig. 1 Schematic of provenance lifecycle

The document goes on to position the definition in the context of Information Management:

> The provenance of information is crucial in deciding whether information is to be trusted, how it should be integrated with other diverse information sources, and how to give credit to its originators when reusing it. In an open and inclusive environment such as the Web, where users find information that is often contradictory or questionable, provenance can help those users to make trust judgements.

1.1 PROV as a Community Data Model and Ontology

The specifications define a data model and an OWL ontology, along with a number of serializations for representing aspects of provenance. The term *provenance*, as understood in these specifications, refers to information about entities, activities, and people involved in producing a piece of data or thing, which can be used to form assessments about its quality, reliability or trustworthiness (PROV-Overview [40]). The specifications include a combination of W3C *Recommendation* and *Note* documents. Recommendation documents include (1) the main PROV data model specification (PROV-DM [34]), with an associated set of constraints and inference rules (PROV-CONSTRAINTS [5]); (2) an OWL ontology that allows a mapping of the data model to RDF (PROV-O [18]), and (3) a notation for PROV with a relational-like syntax, aimed at human consumption (PROV-N [35]). All other documents are Notes. These include PROV-XML, which defines a XSD schema for XML serialization [41]. PROV-AQ, the Provenance Access and Query document [33], which defines a Web-compliant mechanism to associate a dataset to its provenance; PROV-DICTIONARY [39], for expressing the provenance of data collections defined as sets of key-entity pairs; and PROV-DC [38], which provides a mapping between PROV-O and Dublin Core Terms.

1.2 The Provenance of PROV

PROV is the result of a long incubation process within the provenance commnunity, documented for instance in [6]. The idea of a community-grown data model for describing the provenance of data originated around 2006, when consensus began to emerge on the benefits of having a uniform representation for data provenance, process documentation, data derivation, and data annotation, as stated in [32]. The First Provenance Challenge [31] was then launched, to test the hypothesis that heterogeneous systems (mostly in the e-science/cyberinfrastructure space), each individually capable of producing provenance data by observing the execution of data-intensive processes, could successfully exchange such provenance observations with each other, without loss of information. The Open Provenance Model (OPM) [32] was proposed as a common data model for the experiment. Other Provenance Challenges followed, to further test the ability of the OPM to support interoperable provenance.

In September 2009, the W3C Provenance Incubator Group was created. Its mission, as stated in the charter [43], was to "provide a state-of-the art understanding and develop a roadmap in the area of provenance for Semantic Web technologies, development, and possible standardization." W3C Incubator groups produce recommendations on whether a standardization effort is worth undertaking. Led by Yolanda Gil at University of Southern California, the group produced its final report in December 2010 [44]. The report highlighted the importance of provenance for multiple application domains, outlined typical scenarios that would benefit from a rich provenance description, and summarized the state of the art from the literature, as well as in the Web technology available to support tools that exploit a future standard provenance model. As a result, the W3C Provenance Working Group was created in 2011, chaired by Luc Moreau (University of Southampton) and Paul Groth (Vrije Universiteit Amsterdam). The group released its final recommendations for PROV in June 2013.

1.3 Other Notions of Data Provenance

Other formal models of data provenance exist, specifically in the context of database management. The provenance of a data item that is returned by a database query, for example, is defined by the semantics of the query itself, and mentions the fragments of the database state that were involved in the query processing [4]. An algebraic theory in support of data provenance representation and management has been developed [13]. This form of *fine-grained* provenance is often contrasted with *coarse-grained* provenance, which records the input / output derivations that are observed when functions are invoked, typically from within workflows and in the context of scientific data processing [9]. Attempts have also been made to reconcile these two views, e.g., when declarative-style queries are embedded within procedural workflow processing [1].

2 Embracing Provenance: Status and Opportunities

As illustrated in Fig. 1, there are a few key phases in the lifecycle of a provenance document: Production (Capture), persistent storage, Query, Sharing, Association with the underlying data products, and consumption/exploitation (Visualization/-Analysis). The remainder of this short overview will only cover issues concerning Capture, Storage and Query, and Analysis, using the following simple example to illustrate key issues in each of these phases.

In PROV, a provenance document is a set of assertions about the derivations that account for the production of a dataset, including, when available, its attribution. For example, one can use PROV to formally express the following facts:

> Alice took draft v0.1 of paper *P*, made some edits during a certain time interval, and produced a new draft v0.2 of *P*.
> In doing so, she used papers p_1, p_2 as reference.
> Alice then delegated Robert to do proofreading of *P* v0.2, which resulted in a new version v0.3 of *P*.
> Alice also published a dataset *D* as supplementary material to *P*, which she has uploaded to a public data repository, for others to discover and reuse.

These facts can be expressed formally, using either RDF, XML, or PROV-N, the bespoke near-relational syntax mentioned earlier.

2.1 *Extending PROV*

The PROV Working Group worked hard to ensure that PROV can be extended in a principled way, in order to fit the needs of multiple disciplines where expressing the provenance of data may be important. Specifically, one can (1) use PROV-O, the PROV OWL ontology, in conjunction with other ontologies, in order to provide rich semantic annotations of data, and (2) extend PROV-O itself with domain-specific provenance concepts.

As an illustration of (1), in the example above one can semantically characterize data products as "papers" of a certain type, along with the associated activities (editing, proofreading) using a suitable vocabulary, while at the same time characterizing their provenance using an RDF serialisation of the example statements above. As a reference, in the recent past we have demonstrated this capability in our specification of the *Janus* ontology [25]. In brief, provenance and semantic annotations serve complimentary roles: the former tells the *history* of a data product, while the latter elucidates its *meaning*.

Regarding extending PROV, one notable example is the ProvONE ontology (formerly known as D-PROV) [27], aimed at capturing at the same time the data dependencies that emerge from observations during data creation (known as *retrospective* provenance), as well as the static structure of the process that is responsible for the generation of the process (known as *prospective* provenance)

[21]. The latter is deliberately missing from PROV, owing to its generality. The D-PROV ontology specifically extends PROV to account for the structure of scientific workflows, a specific type of data-generating process that is important in many e-science applications.

In particular, the latest embodiment of D-PROV, called ProvONE [45], is currently in production use by the DataONE project (dataone.org). DataONE, a large NSF-funded project (2010–2018), is the largest Research Data conservancy project in the USA, with a focus on Earth Observational Data and ecology/climate data in particular. With a growing federation that already counts tens of member repositories and hundreds of thousands of science data objects, the DataONE architecture places metadata indexing and management at the cornerstone of its data search and discovery capabilities. "Searching by provenance" is a new and unique feature that leverages the ProvONE data model, as well as the automated capture of retrospective provenance whenever R or Matlab (and, soon, Python) scripts that access DataONE science objects are executed.

The ProvONE ontology provides a template for extending PROV, which can be used in a number of other domains, as it illustrates proper use of the PROV extensibility points.

2.2 *Provenance Capture*

Provenance is the result of observing a data transformation process in execution, including details of its inputs and outputs, be it a database query or a workflow, including processes carried out by humans or only partially automated. Key questions concerning the recording ("capturing") of provenance include (1) what provenance-related events can be observed, (2) what is the level of detail of these observations, and (3) how does one deal with multiple, overlapping but inconsistent observations?

Regarding scientific data processing, the ability to record provenance relies entirely on the infrastructure on which the processes are executed. An increasing number of tools and systems are being retrofitted with provenance recording capabilities, including the best known workflow management systems [9, 28], and more recently, the Python [37] and the R languages [19, 20] for data analytics. Two specific instances of provenance capture sub-systems for scientific workflows, that we have actively contributed to, are [24], for the Taverna workflow management system developed in Manchester to support bioinformatics researchers [15, 26], and for the eScience Central workflow manager [14].

The case of completely automated processes which run in a centralized environment is, however, the simplest possible scenario. "Human-in-the-loop" processes are obviously more problematic, and are limited to capturing human interactions with information systems through a user interface. Clearly, solutions in this space are necessarily bespoke, with no known publications reporting specific case studies.

In each of these cases, the observations may be available at a specific level of abstraction, which may or may not be appropriate for the type of downstream analysis requirements (see below). These range from fine-grained, high-volume, system-level provenance (ie every file I/O operation in the system) [23], to "coarse-grained" provenance from workflow executions, where only the inputs and outputs of each workflow block can be observed.

As a consequence of these varying levels of details, it becomes necessary to be able to adjust the quantity of information contained in a provenance document, i.e., by creating *views over provenance* that represent abstractions over provenance. In the example above, we could for instance conflate the editing and proofreading activities into one, high-level "paper preparation" activity, and ignore the interim v0.2 of *P*. Our own work on *provenance abstraction* [29] builds upon prior research [2, 10], reflecting the user need not only to simplify the amount of provenance presented to the used, but also to *obfuscate* provenance in order to preserve its confidentiality.

A further complication in provenance capture, is that the observable processes normally take place on multiple, heterogeneous, autonomous and distributed systems, where the corresponding data is scattered. The provenance of an end data product must therefore be *reconstructed* by composing multiple, possibly inconsistent, and incomplete provenance fragments harvested from each of those systems. This is a relevant but under-studied area of research for provenance, with many potential applications that extend well beyond the realm of e-science.

2.3 Storage, Retrieval, and Query

Storing, indexing, and querying provenance documents requires a data layer not unlike that used to store the underlying data products that the provenance refers to. Data provenance that describes the history of large volumes of data is itself bound to have a high volume. Furthermore, if one includes in the provenance the intermediate data products that are generated as part of a complex data processing pipeline, it is easy to see that the size of the provenance documents may vastly exceed that of the data whose history it describes. Older and recent research has been devoted to studying the trade-offs between storing intermediate data products as part of provenance, which may incur a high storage cost [46], as opposed to partially re-computing the data products (" smart rerun" [7]).

Issues of dealing with large-scale provenance were addressed in the *BigProv* international workshop organized in 2013 and co-located with the EDBT conference. A number of submissions contributed to corroborate the hypothesis that the scalability of provenance management systems is becoming a practical problem if interesting analytics are to be derived from it. Amongst these, a study on reconstructing provenance from log files [12].

Provenance documents such as the one in our example are naturally expressed in the form of a graph. This suggests that graph databases (GDBMS) are suitable for their persistent storage, indexing, and querying. In our past work we have been experimenting with Neo4J, a new generation GDBMS, in order to study the scalability properties of provenance storage. In particular, we have developed ProvGen [11], a generator of synthetic provenance graphs of arbitrary size and with topology constraints. ProvGen is designed to create benchmarks for testing the performance of graph-based provenance data layers. It can generate provenance documents with millions of nodes and stores them in a Neo4J database.

At the same time, the standard RDF serialization of PROV, which specifies how provenance documents can be expressed using RDF triples that comply with the PROV ontology (PROV-O), lends itself well to storing provenance graphs in existing RDF triple stores. However, despite the need for testing provenance data layers at scale, and our own past attempts at soliciting contributions that document scalability of provenance storage and query systems (the ProvBench workshop, co-located with *BigProv* (see above), to the best of our knowledge no official benchmarks have ever been released.

2.4 Provenance Analytics and Novel Uses for Provenance

With the broad term "provenance analytics" we indicate all forms of consumption and exploitation of provenance corpora, once they have been captured and made available through suitable data engineering solutions, alluded to above. Relevant questions include: what can we learn from a large body of provenance metadata? what techniques and algorithms can be successfully borrowed from the realm of (Big) Data Analytics, in order to gain insight into data through its provenance?

Much has been made of provenance as a key form of metadata to help understanding the quality of data as well as its trustworthiness. A whole special issue of the ACM Journal of Data and Information Quality, has been devoted to the topic [42]. Despite several high quality submissions, however, more research is needed to fully elucidate the connection between data provenance and quality.

Many other opportunities are worth exploring that exploit provenance corpora in several domains. One line of research still in its infancy, concerns using provenance to ascribe *transitive credit* [16] to scientists and other contributors who publish their datasets in public data repositories, for others to reuse. Data publication is a rapidly growing area of Open Science, which is based upon the assumption that scientists will spontaneously make their datasets public, as long as due credit is given to them through community mechanisms. Unfortunately, these mechanisms are still quite primitive, limited as they are to counting the number of citations to datasets, as they are found in paper publications (see for instance the *Making Data Count* project [17]). Instead, transitive credit pushes this embrionic notion of "credit for data" much further, as it leverages provenance to take into account multiple generations of data derivation and reuse.

Other disciplines farther away from computing and science will benefit from properly collected provenance, wherever providing accountability of a process execution is important. One example amongst many concerns *food safety*, where traceability of lots of food along a supply chain is critical to ensuring compliance with quality standards and proper handling, and to answer questions in case of accidents involving consumption of unsafe food.

2.5 Three Key Challenges for Practical Usability of Provenance Data

To conclude this overview, three areas when more research is needed in order to make provenance usable in practice are worth mentioning.

Incomplete and Uncertain Provenance Generation and usage of data naturally occurs in many different ways through multiple, autonomous information systems. As a consequence, the provenance of such data is also naturally *fragmented and incomplete*. One major problem in provenance research is how to reconstruct a complete "big picture" out of such fragments. We are currently addressing this foundational problem in the specific setting of Open Research Data reuse, as this is a key issue when establishing transitive credit as mentioned above.

Trusted Provenance A second issue concerns accountability of the provenance documents themselves. To the extent that provenance documents are considered as a form of evidence for the underlying data, it is necessary to ensure that the provenance itself can be trusted not to have been tampered with. Using provenance traces in, say, a court of law, requires strong non-repudiability and integrity guarantees, which can only be provided by a trusted computing infrastructure [22, 30]. The notion of tamper-proof (or rather, tamper-evident) provenance has been touched upon in the past [47], but more research is needed as this clearly conflicts with the notion of provenance abstraction through views, alluded to above, namely when generating views involves *redacting* the provenance document itself [3].

Provenance to Help the Reproducibility of Scientific Processes Lastly, we mention a long-standing promise on which provenance studies have largely yet to deliver. Much has been said (and there is no scope for a full survey here) of the role of provenance to support reproducible science, since the connection between reproducibility and provenance was first made back in 2008 [8].

Reproducibility is a known problem for a large number of scientific processes of the past, which are often encoded as a loose collection of scripts with external dependencies on ever-changing libraries, services, and databases. Practical solutions where provenance is used to ensure that these processes are reproducible are not readily available, however. In the recent past, we have addressed one aspect of this problem, namely by showing that provenance traces can be used to explain the differences between two sets of results that are obtained from the executions of two versions of a process [28], the latest being a reproduction of the original.

Much remains to be done, however, to clearly prove the role of provenance data in data-driven, reproducible science.

References

1. Amsterdamer, Y., Davidson, S.B., Deutch, D., Milo, T., Stoyanovich, J., Tannen, V.: Putting lipstick on pig: enabling database-style workflow provenance. Proc. VLDB Endow. **5**(4), 346–357 (2011)
2. Biton, O., Cohen-Boulakia, S., Davidson, S.B.: Zoom*UserViews: querying relevant provenance in workflow systems. In: VLDB, pp. 1366–1369 (2007)
3. Cadenhead, T., Khadilkar, V., Kantarcioglu, M., Thuraisingham, B.: Transforming provenance using redaction. In: Proceedings of the 16th ACM Symposium on Access Control Models and Technologies, SACMAT '11, pp. 93–102. ACM, New York (2011)
4. Cheney, J., Chiticariu, L., Tan, W.-C.: Provenance in databases: why, how, and where. Found. Trends Databases **1**, 379–474 (2009)
5. Cheney, J., Missier, P., Moreau, L.: Constraints of the provenance data model. Technical Report (2012)
6. Cheney, J., Finkelstein, A., Ludaescher, B., Vansummeren, S.: Principles of provenance (Dagstuhl Seminar 12091). Dagstuhl Reports **2**(2), 84–113 (2012)
7. Cohen-Boulakia, S., Leser, U.: Search, adapt, and reuse: the future of scientific workflows. SIGMOD Rec. **40**(2), 6–16 (2011)
8. Davidson, S., Freire, J.: Provenance and scientific workflows: challenges and opportunities. In: Proceedings of SIGMOD Conference, Tutorial, pp. 1345–1350 (2008)
9. Davidson, S., Cohen-Boulakia, S., Eyal, A., Ludäscher, B., McPhillips, T., Bowers, S., Anand, M.K., Freire, J.: Provenance in scientific workflow systems. In: Data Engineering Bulletin, vol. 30. IEEE, New York (2007)
10. Dey, S., Zinn, D., Ludäscher, B.: ProPub: towards a declarative approach for publishing customized, policy-aware provenance. In: Cushing, J.B., French, J., Bowers, S. (Eds.), Scientific and Statistical Database Management. Lecture Notes in Computer Science, vol. 6809, pp. 225–243. Springer, Berlin, Heidelberg (2011)
11. Firth, H., Missier, P.: ProvGen: generating synthetic PROV graphs with predictable structure. In: Proceedings of IPAW 2014 (Provenance and Annotations), Koln (2014)
12. Ghoshal, D., Plale, B.: Provenance from log files: a bigdata problem. In: Proceedings of BigProv Workshop on Managing and Querying Provenance at Scale (2013)
13. Green, T.J., Karvounarakis, G., Tannen, V.: Provenance semirings. In: PODS, pp. 31–40 (2007)
14. Hiden, H., Watson, P., Woodman, S., Leahy, D.: e-Science central: cloud-based e-Science and its application to chemical property modelling. Technical Report cs-tr-1227. School of Computing Science, Newcastle University (2011)
15. Hull, D., Wolstencroft, K., Stevens, R., Goble, C.A., Pocock, M.R., Li, P., Oinn, T.: Taverna: a tool for building and running workflows of services. Nucleic Acids Res. **34**, 729–732 (2006)
16. Katz, D.S.: Transitive credit as a means to address social and technological concerns stemming from citation and attribution of digital products. J. Open Res. Soft. **2**(1), e20 (2014)
17. Kratz, J.E., Strasser, C.: Making data count. Nature Scientific Data **2**, 150039 (2015)
18. Lebo, T., Sahoo, S., McGuinness, D., Belhajjame, K., Cheney, J., Corsar, D., Garijo, D., Soiland-Reyes, S., Zednik, S., Zhao, J.: PROV-O: The PROV ontology. Technical Report (2012)
19. Lerner, B.S., Boose, E.R.: Collecting provenance in an interactive scripting environment. In: Proceedings of TAPP'14 (2014)
20. Lerner, B., Boose, E.: RDataTracker: collecting provenance in an interactive scripting environment. In: 6th USENIX Workshop on the Theory and Practice of Provenance (TaPP 2014) (2014)

21. Lim, C., Lu, S., Chebotko, A., Fotouhi, F.: Prospective and retrospective provenance collection in scientific workflow environments. In: 2010 IEEE International Conference on Services Computing (SCC), pp. 449–456 (2010)
22. Lyle, J., Martin, A.: Trusted computing and provenance: better together. In: Proceedings of the 2nd Conference on Theory and Practice of Provenance, TAPP'10, Berkeley, CA, p. 1. USENIX Association, Berkeley, CA (2010)
23. Macko, P., Chiarini, M., Seltzer, M.: Collecting provenance via the Xen hypervisor. In: Freire, J., Buneman, P. (eds.) TAPP Workshop, Heraklion (2011)
24. Missier, P., Paton, N., Belhajjame, K.: Fine-grained and efficient lineage querying of collection-based workflow provenance. In: Proceedings of EDBT, Lausanne, Switzerland (2010)
25. Missier, P., Sahoo, S.S., Zhao, J., Sheth, A., Goble, C.: Janus: from workflows to semantic provenance and linked open data. In: Proceedings of IPAW 2010, Troy, NY (2010)
26. Missier, P., Soiland-Reyes, S., Owen, S., Tan, W., Nenadic, A., Dunlop, I., Williams, A., Oinn, T., Goble, C.: Taverna, reloaded. In: Gertz, M., Hey, T., Ludaescher, B. (eds.) Proceedings of SSDBM 2010, Heidelberg (2010)
27. Missier, P., Dey, S., Belhajjame, K., Cuevas, V., Ludaescher, B.: D-PROV: extending the PROV provenance model with workflow structure. In: Proceedings of TAPP'13, Lombard, IL (2013)
28. Missier, P., Woodman, S., Hiden, H., Watson, P.: Provenance and data differencing for workflow reproducibility analysis. Concurr. Comput. **28**(4), 995–1015 (2016)
29. Missier, P., Bryans, J., Gamble, C., Curcin, V., Danger, R.: ProvAbs: model, policy, and tooling for abstracting PROV graphs. In: Proceedings of IPAW 2014 (Provenance and Annotations), Koln. Springer, Berlin (2014)
30. Mitchell, C., Mitchell, C., Mitchell, C.: Trusted computing. In: Chen, L., Mitchell, C.J., Martin, A. (eds.) Proceedings of Trust 2009, Oxford. Springer, Berlin (2005)
31. Moreau, L., Ludäscher, B., Altintas, I., Barga, R.S.: The first provenance challenge. Concurr. Comput. **20**, 409–418 (2008)
32. Moreau, L., Clifford, B., Freire, J., Futrelle, J., Gil, Y., Groth, P., Kwasnikowska, N., Miles, S., Missier, P., Myers, J., Plale, B., Simmhan, Y., Stephan, E., Van Den Bussche, J.: The open provenance model—core specification (v1.1). Futur. Gener. Comput. Syst. **7**(21), 743–756 (2011)
33. Moreau, L., Hartig, O., Simmhan, Y., Myers, J., Lebo, T., Belhajjame, K., Miles, S.: PROV-AQ: provenance access and query. Technical Report (2012)
34. Moreau, L., Missier, P., Belhajjame, K., B'Far, R., Cheney, J., Coppens, S., Cresswell, S., Gil, Y., Groth, P., Klyne, G., Lebo, T., McCusker, J., Miles, S., Myers, J., Sahoo, S., Tilmes, C.: PROV-DM: the PROV data model. Technical Report. World Wide Web Consortium (2012)
35. Moreau, L., Missier, P., Cheney, J., Soiland-Reyes, S.: PROV-N: the provenance notation. Technical Report (2012)
36. Moreau, L., Groth, P., Cheney, J., Lebo, T., Miles, S.: The rationale of PROV. Web Semant. Sci. Serv. Agents World Wide Web **35**, Part 4, 235–257 (2015)
37. Murta, L., Braganholo, V., Chirigati, F., Koop, D., Freire, J.: noWorkflow: capturing and analyzing provenance of scripts. In: Proceedings of IPAW'14 (2014)
38. PROV DC (2013). Available at http://www.w3.org/TR/prov-dc/
39. PROV Dictionary (2013). Available at http://www.w3.org/TR/prov-dictionary/
40. PROV-Overview: An Overview of the PROV Family of Documents. Technical Report (2012)
41. PROV-XML (2013). Available at http://www.w3.org/TR/prov-xml/
42. Special Issue on Provenance, Data and Information Quality. J. Data Inf. Qual. **5**(3) (2015)
43. The Provenance Incubator Group Charter (2009). Available at http://www.w3.org/2005/Incubator/prov/charter
44. The Provenance Incubator Group Final Report (2010). Available at http://www.w3.org/2005/Incubator/prov/XGR-prov-20101214/
45. The ProvONE provenance model (2014). Available at http://tinyurl.com/ProvONE

46. Woodman, S., Hiden, H., Watson, P.: Workflow provenance: an analysis of long term storage costs. In: Proceedings of 10th WORKS workshop, Austin, TX (2015)
47. Zhang, J., Chapman, A., LeFevre, K.: Do you know where your datas been? tamper-evident database provenance. In: Jonker, W., Petkovic, M. (eds.) Secure Data Management. Lecture Notes in Computer Science, vol. 5776, pp. 17–32. Springer, Berlin/Heidelberg (2009)

Part V
Cognitive Science Perspectives Through the Lens of Visual Analytics

Visual Analytics: Data, Analytical and Reasoning Provenance

Margaret Varga and Caroline Varga

Abstract Analysts and decision makers are increasingly overloaded with vast amounts of data/information which are often dynamic, complex, disparate, conflicting, incomplete and, at times, uncertain. Furthermore, problems and tasks that require their attention can be ambiguous, i.e. they are ill-defined. In order to make sense of complex data and situations and make informed decisions, they utilize their intuition, knowledge and experience. Provenance is fundamental for the user to capture and exploit effectively the explicit data and implicit knowledge within the decision making process. Provenance can usefully be considered at three conceptual levels, namely: data (what), analytical (how) and reasoning (why). This paper explores visual analytics in the exploitation of provenance within the decision making process.

Keywords Analytical provenance • Data provenance • Hypothesis • Reasoning provenance • Visual analytics • Visualization

1 Introduction

Analysts and decision makers are increasingly overloaded with vast amounts of data/information which are often dynamic, complex, disparate, conflicting, incomplete and, at times, uncertain. Furthermore, problems and tasks that require their attention can be ambiguous, i.e. they are *ill-defined*. In order to make sense of complex data and situations and make informed decisions, decision makers rely on explicit information and their implicit intuition, knowledge and experience. Moreover, to have confidence in a decision making process, it is necessary for them

M. Varga
Seetru Ltd, Bristol, UK,
e-mail: margaret.varga@seetru.com; margaret.varga@oncology.ox.ac.uk

University of Oxford, Oxford, UK

C. Varga
Seetru Ltd, Bristol, UK

V.L. Lemieux (ed.), *Building Trust in Information*, Springer Proceedings in Business and Economics, DOI 10.1007/978-3-319-40226-0_9

to understand the sources of information and thus the value and trust that can be placed on every aspect of the process; i.e. the provenance [1–4].

Provenance is fundamental for the user to capture and exploit effectively the explicit data and implicit knowledge within a decision making process. Provenance can usefully be considered at three conceptual levels, namely: data, analysis and reasoning [5]. In essence it comprises the what (data), how (it was analyzed) and why (reasoning).

This paper explores the application of visual analytics as an effective means of analyzing and understanding provenance in the explicit representation of the analytical and reasoning processes: how and why the data is used.

2 Data, Analytical and Reasoning Provenance

There are three categories of provenance that play a role in visual analytics, namely: data provenance, analytical provenance and reasoning provenance. In order to understand findings/discoveries it is necessary to document the entire analysis process and retain all three types of provenance. Capturing the reasoning processes is by far the most challenging.

- **Data provenance** considers the source of the data, and the link between the source and the system using the data. The data may be intelligence reports, videos, network logs, etc. The provenance of the data must certainly be taken into account in the analytical and visualization approaches and processes when addressing problems/making decisions.
- **Analytical provenance** is concerned with the processes performed on the data; in particular, here, the techniques used to analyze and visualize the data. The analysis conducted has an impact on the nature of the results and how the results can be used. The actions performed during an analysis within a visual analytic system can be captured: i.e. data transformations, events (e.g. key strokes) and actions (e.g. zoom) can easily be logged, and the overall history of interactions can be recorded [4]. VisTrails, for example, supports exploratory computational tasks and also provides a provenance management infrastructure [6].
- **Reasoning provenance** is the most challenging to identify, make explicit and capture; it is concerned with how and why analysts arrive at their conclusions/decisions. It is typically concerned with the application of human experience, knowledge and intuition.

Annotation of the analysis can be used to enable recall and sharing [7]. Externalization can be achieved through think-aloud protocols: this process, however, may alter the nature of the reasoning, reduce task performance, or even risk changing decisions [8, 9]. Furthermore, analysis of such externally captured data is extremely time-consuming and labour intensive.

The results of experience, knowledge and intuition used in a decision making process can be presented in visual (e.g. diagrams), textual or numeric narratives.

These narratives are based on the available data and are used to show salient information and different hypotheses. In the reasoning process, interconnections between narratives, and between information and hypotheses, are developed to support informed decision-making [5].

Visual representations of the reasoning space through networks of narratives enable the understanding of the reasoning process and considerably improve both the quality of the reasoning process and the efficiency/effectiveness of informed decision making [5, 10, 11].

3 Visual Analytics

Visual Analytics is the science of analytical reasoning facilitated by interactive visual interfaces. It combines automated analysis techniques with interactive visualizations to allow the user to interact with, explore and analyze big and complex data, both dynamically and visually. It thus facilitates data and situational understanding so as to support informed decision-making.

It is necessary to create tools and techniques to help users to derive information and insight from massive and complex data: to detect the expected and/or discover the unexpected. The tools must also support the provision and communication of timely and accurate situation assessments—upon which users can act [11–14].

4 Intelligence Analysis

4.1 Introduction

Across all subject domains, one concern is how to incorporate and make use of provenance to enhance informed decision making—to better understand how and why data is used and decisions are made. This section uses a case study on intelligence analysis to illustrate the ideas.

Intelligence analysis is the application of individual and collective cognitive methods to explore data and test hypotheses. Events and evidence are assessed, for example, to explain/interpret events that 'might' happen; or to decide how best to prevent the occurrence of an adverse event; or to minimize potential damage [15, 16], etc.

Intelligence analysts respond to Intelligence Requests (IRs), which can be precise or ill-defined. Critical thinking is essential in order to provide the 'best possible' answer, within a short time frame. Important elements of critical thinking are to reduce bias and present all possible options to a decision maker [17].

4.2 An Example Case Study

An example case study of an ill-defined IR is discussed here: the IR is—"What are the current threats?". The threats may be of any form and of varying degrees of urgency.

In order to determine potential threats data/information must be gathered and analyzed, and the situation must be assessed. The experience of an intelligence analyst will guide them to filter 'noisy' data and to zoom into potential threats that require further investigation. If it is known, for example, that there is an upcoming state visit or a military supply convoy, the analysts' experience/intuition might lead them to identify such events as potential targets, and thence to hypothesize possible/likely threats. Any of the hypothesized threats may be true so analysts must consider all relevant information to make informed decisions.

For example, in the case of a convoy, an ambush may be identified as the mostly likely threat [18]; the analyst must then hypothesize likely ambush locations and gather evidence to answer question such as:

- Where and when is the convoy going, and what route will it take?
- Where have recent ambushes been?
- Where are insurgents currently known to be operating?
- What types of ambushes are the insurgents capable of? Land or sea? Chemical or biological?
- What is the certainty that this route is not going to change (e.g. commander deciding to change the route, flooding, unexpected roadblock, etc . . .)?
- What is the weather forecast? Hurricane? Snowing?
- . . . .

Many different approaches may be used to narrate, assimilate and analyse hypotheses and evidence. Here, the Wigmore concept [19] is used to demonstrate the generation, representation and analysis of multiple hypotheses, as well as provide an answer to the IR including the representation/presentation of the analytical provenance [18]. The Wigmore chart was developed as a graphical method for the analysis of legal evidence in trials. It was the first diagrammatic system of charting arguments; other approaches include Toulmin [20]. One of the advantages of the Wigmore approach is its handling of the balance of view (cf. bias). In a Wigmore chart, various types and items of evidence supporting and refuting a hypothesis are represented graphically; this allows the strength/weakness of the case to be readily observed. In particular, it is easy to see 'gaps' where additional effort is required to gather evidence, e.g. to minimise uncertainty or danger of self-confirmation, or to strengthen an aspect of the case/hypothesis.

In Fig. 1, the hypothesis introduced suggests that a potential ambush location is south of Village A [18]. The analyst enters hypothesis properties based on their experience/intuition/knowledge and their information sources. Sources are rated in terms of their reliability [15]:

(1) Completely reliable,

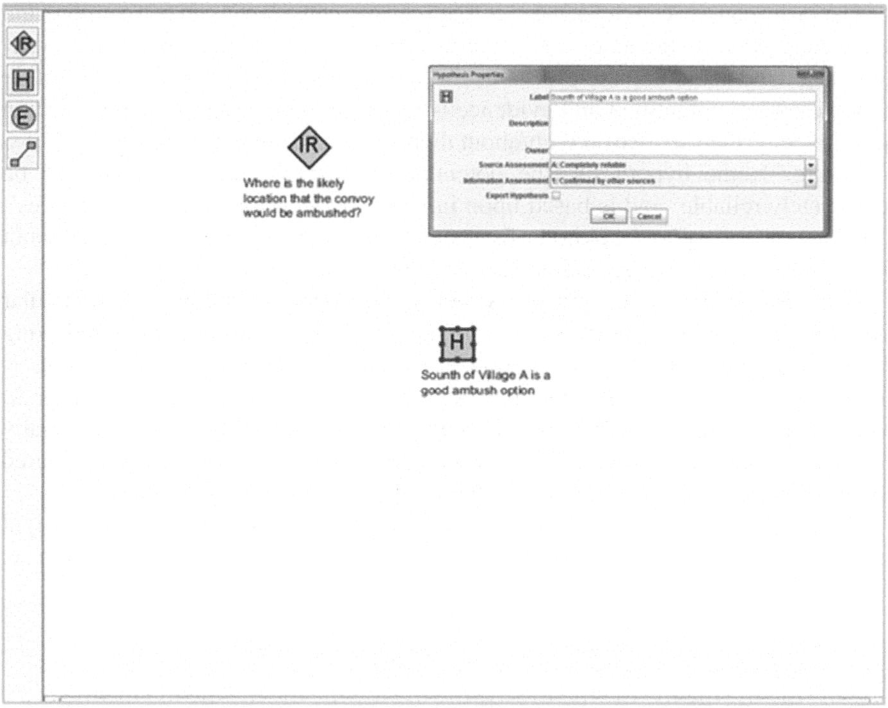

Fig. 1 Hypothesis that South of Village A is a good ambush option and its properties [18]

(2) Usually reliable,
(3) Fairly reliable,
(4) Not usually reliable,
(5) Unreliable, and,
(6) Reliability cannot be judged.

The information is also rated accordingly; see again Fig. 1:

(1) Confirmed by other sources,
(2) Probably true,
(3) Possibly true,
(4) Doubtful,
(5) Improbable, and,
(6) Truth cannot be judged.

In this way, multiple ambush locations (hypotheses) based on the analysts' experience—obvious and less obvious alternatives—can be considered, as opposed to pursuing a biased approach in which, for example, a narrow focus might be pursued.

The provenance of each hypothesis that is being considered can be recorded and assessed; guided by the analyst's intuition, experience, local knowledge and available data. Understanding the processes through which ambush locations—or other hypotheses—are identified and evidence is assessed is important in understanding how and why analysts reason with/about them (i.e. reasoning provenance).

In Fig. 1, the hypothesis (the potential location) source is assessed to be 'completely reliable' and is based upon information 'confirmed by other sources'. The Analyst can export the hypothesis to share with other analysts, make notes about the hypothesis (reasoning), and declare any ownership.

Next, the analyst must gather evidence to support or refute this particular hypothesis (i.e. build a balance of view). The evidence properties are entered using the same rating system as the hypothesis properties. In this example, there is supporting evidence to show that there is good cover for the attackers at Village A, which makes Village A vulnerable. This evidence is believed to be from a 'usually reliable' source and is 'probably true', see Fig. 2. More evidence will be gathered and similar processes will be used for other evidence and other hypotheses.

Different types of evidence can be used; for example, significant trends might emerge from circumstantial evidence that can be correlated with other evidence.

Fig. 2 Properties of the evidence that there is a good cover for the attackers [18]

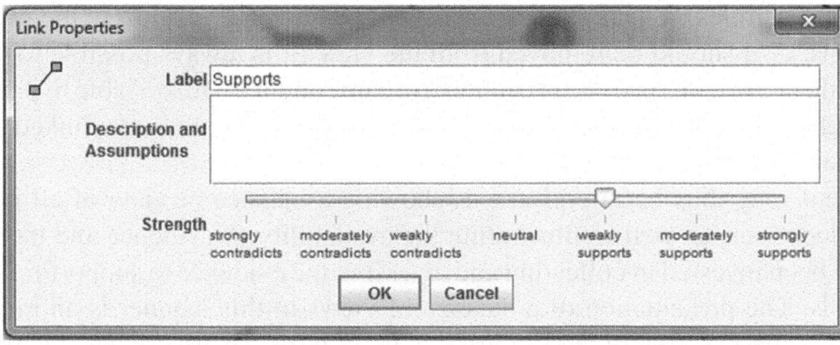

Fig. 3 Link properties [18]

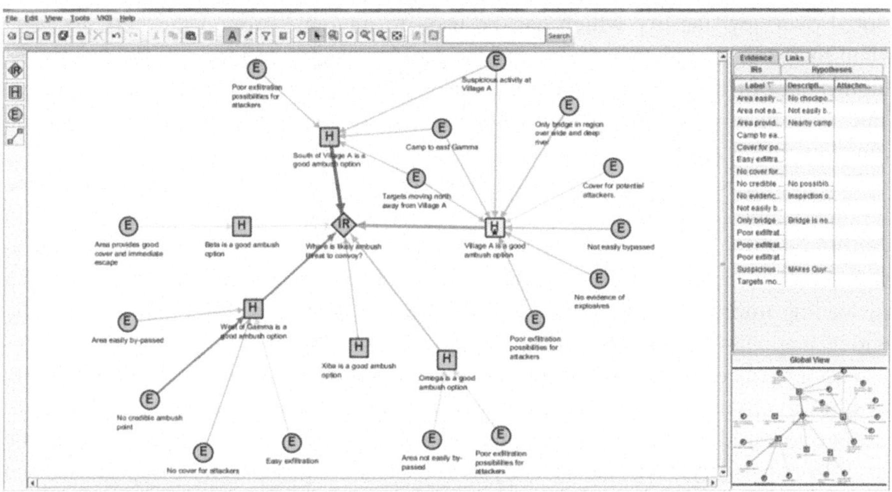

Fig. 4 Network of six hypotheses [18]

The 'Is Pending', checkbox allows the analyst to represent evidence that someone has been tasked to collect, see Fig. 2. This can be added to the graph to be assessed later when the evidence is available. The origin of the evidence is also recorded; that is, its data provenance.

The analyst also enters the strength of the evidence in supporting or refuting the hypothesis; this ranges from strongly contradicts, neutral, to strongly supports, see Fig. 3. This rating is used to determine the role each item of evidence plays in supporting/refuting the hypothesis (it should be noted that the reliability of a source does not determine its relevance or strength).

Figure 4 shows a network of hypotheses (denoted 'H') and evidence (denoted 'E') built up in answering the IR. Green links represent support while red links represent refutation; their strength is represented by the thickness of the link.

Among the six possible locations/hypotheses, no evidence has been collected for the hypothesis that Xiba is a possible ambush location (there are no linked evidence/'E' circles): either evidence needs to be gathered to address the proposed hypothesis, or it should be removed from the view (it is always possible to display the hypothesis again if evidence becomes available and it is still a viable hypothesis). Conversely, many items of evidence—of varying strengths—are linked to the hypothesis ('H' box) of Village A.

Figure 4 may thus be interpreted as showing a balance of view of all possible ambush locations, as well as illustrating the availability of evidence and the efforts that have been invested in collecting and assessing the evidence to support/refute any hypothesis. The presentation of a balance of views in this manner is an important element in reducing cognitive bias. Inputs from other analysts can also reduce bias.

The upper right hand table on Fig. 4 provides a means by which all evidence and associated information may be examined. The bottom right hand panel provides a global overview facility which can be used to navigate around the network: e.g. when focusing on sub-components of the problem such as Village A. Alternatively, another part of the network may be looking at the threats relating to an imminent state visit; in this case the global overview would show all the possible threats and how they might relate.

The same evidence may be used to support multiple hypotheses, or to support one hypothesis and refute another; it is not necessary to input the same evidence again for different hypotheses. For example, "suspicious activity at Village A" relates to multiple hypotheses, see Fig. 4. The advantage of this is that it avoids creating a misleading impression of the number of evidence items available; that is, the same evidence appearing multiple times for different hypotheses.

The influence of individual and combined evidence is analyzed automatically for the six hypotheses. It is vital that the system can readily be updated to respond to rapidly evolving situations; here, the effect of new information can be visualized instantly. Hypothesis analysis is a dynamic process; new hypotheses can be generated or removed when the situation changes, when new evidence is gathered or when there are changes in existing evidence/situations. Hypotheses can also be saved, re-used or modified for future IRs.

Different analytical processes can be used to analyze the data. In this example, the system calculates the strength of all the evidence relating to all the hypotheses; this reveals that Village A is indeed a likely ambush location. In light of this, the convoy should either alter its route to avoid Village A or prepare for a possible ambush. The display can also be used to brief the commander about all the threats considered, which is the most likely threat, and why. The understanding of the hypotheses and corresponding evidence as well as the analysts' notes give an idea of the reasoning of why the locations were chosen.

This concept can be transferred for use in other applications such as financial risk analysis or medical analysis [18].

5 Conclusions

The case study has illustrated the interactive formation, visualization and analysis of dynamic hypotheses for decision-making in ill-defined problems; with provenance for explicit data, defined analytical process and some implicit knowledge/experience from the analysts. It shows the benefits of interactive analysis and visualization in response to changes in evidence and hypotheses; such as reducing bias and improving efficiency. Furthermore, the approach considers the degree of uncertainty in each piece of evidence and its role in supporting/refuting different hypotheses. The data provenance and the data uncertainty can be expressed by the user based on assessment of its source, assessment of the information, as well as consideration of the links between the evidence and hypotheses. It also provides an audit trail of the analysis in terms of data provenance, analytical provenance, and, to certain extent, reasoning provenance.

This case study shows narratives as the explicit representations of the hypotheses, which include different types of data in their presentation, such as: the explicit data used and its provenance; the processing and manipulation performed; and, the implicit information from the analysts' knowledge and experience.

Systems that allow for the dynamic visualization of hypotheses which develop over time, and change with the arrival of 'new information' or the application of a 'new process', provide invaluable support for informed and dynamic decision making in ill-defined problems. The system in the illustrated case study is an example of this capability. It also provides methods to visualize competing hypotheses or complementary theories (that would support and enhance the strength of a particular argument), each depicting different degrees of certainty.

The case study shows that although many pieces of the puzzle have been found, much research is still needed to further the development of tools to support informed decision making for ill-defined problems. Robust reasoning provenance about how and why analysts make decisions, deduced from implicit data, would complete the audit trail for understanding what, how and why data were used.

References

1. Attfield, S.J., Hara, S.K., Wong, B.L.W.: Sensemaking in visual analytics: processes and challenges. In: Kohlhammer, J., Keim, D. (eds.) EuroVAST 2010: International Symposium on VAST, pp. 1–6. Eurographics Association, Bordeaux, France (2010)
2. Gotz, D., Zhou, M.X.: Characterizing users' visual analytic activity for insight provenance. Inf. Vis. **8**(1), 42–55 (2009)
3. Jankun-Kelly, T.J.: The Case for Visual Analysis Provenance Cases, Workshop on Analytic Provenance: Process + Interaction + Insight, CHI. (2011)
4. Venters, C.C., Austin, J., Dibsdale, C.E., Dimitrova, V., Djemame, K., Fletcher, M., Fores, S., Hobson, S., Lau, L., McAvoy, J., Marshall, A., Townend, P., Taylor, N., Viduto, V., Webster, D.E., Xu, J.: To trust or not to trust? Developing trusted digital spaces through timely reliable and personalized provenance. In: Provenance for Sensemaking. Paris, France (10th November 2014)

5. Roberts, J.C., Keim, D., Hanratty, T., Rowlingson, R., Hall, M., Jacobson, Z., Lavigne, V., Rooney, C., Varga, M.: From Ill-defined Problems to Informed Decisions. EuroVis Workshop on Visual Analytics, UK (2014)
6. Silva, C.T., Freire, J., Callahan, S.: Provenance for visualizations: reproducibility and beyond. Comput. Sci. Eng. **9**(5), 82–90 (2007)
7. Groth, D., Streefkerk, K.: Provenance and annotation for visual exploration systems. IEEE Trans. Vis. Comput. Graph. **12**(6), 1500–1510 (2006)
8. Boren, T., Ramey, J.: Thinking aloud: reconciling theory and practice. IEEE Trans. Prof. Commun. **43**(3), 261–278 (2000)
9. Hertzum, M., Hansen, K.D., Andersen, H.H.K.: Scrutinising usability evaluation: does thinking aloud affect behavior and mental workload? Behav. Inf. Technol. **28**(2), 165–181 (2009)
10. Schacter, D.: Psychology, 2nd edn. Worth Publishers, NY (2009)
11. Thomas, J.J., Cook, K.A.: Illuminating the Path: The Research and Development Agenda for Visual Analytics. National Visualization and Analytics Centre (2005)
12. Keim, D.A., Kohlhammer, J., Ellis, G., Mansmann, F.: Mastering the Information Age-Solving Problems with Visual Analytics. Florian Mansmann (2010)
13. Keim, D.A., Mansmann, F., Thomas, J.: Visual analytics: how much visualization and how much analytics? SIGKDD Explor. **11**(2), 5–8 (2009)
14. Thomas, J.J.: Taxonomy for Visual Analytics: Seeking Feedback. VAC Views (May 2009)
15. FM 2-22.3 (FM 34-52) Human Intelligence Collector Operations, US Department of the Army (September 2006)
16. Johnson, R.: Analytics Culture in the U.S. Intelligence Community: An Ethnographic Study, Centre for the Study of Intelligence. Central Intelligence Agency, Washington (2005)
17. Moore, D.T.: Critical Thinking and Intelligence Analysis, Occasional Paper Number 14. National Defense Intelligence College, Washington (March 2007)
18. Varga, M.J., Adams, K.: Interactive hypothesis visualization. In: NATO Workshop on Visualising Networks: Coping with Change and Uncertainty (October 2010)
19. Anderson, T., Schum, D., Twining, W.: Analysis of Evidence, 2nd edn. Cambridge University Press (2005)
20. Toulmin, S.E.: The Uses of Argument - Updated Edition. Cambridge University Press, Cambridge (2003)

Analytic Provenance and Distributed Sensemaking

Ashley Wheat, Simon Attfield, and Robert Fields

Abstract Analytic provenance is a record of reasoning over time, accounting for the methods and techniques used. In sensemaking—where people embark on a process of comprehension by which they gain meaning and insight from information—a record of provenance can support the scrutiny of findings, reflection on the reasoning process, and handover of tasks in collaborative settings. However, sensemaking does not occur within a vacuum, and often involves use of various representational media and artifacts such as maps, charts and lists to gain insight. Therefore, a complete account of analytic provenance in sensemaking scenarios must include descriptions of the use of these representational media. In this paper we discuss analytic provenance in the context of *distributed sensemaking*, showing how we can model the use of representational artifacts and reasoning over time as *inference trajectories*, introduce levels of description of representational artifacts and discuss challenges faced in the capture of analytic provenance in distributed sensemaking scenarios.

1 Introduction

Sensemaking refers to a process of comprehension by which human beings formulate a plausible understanding and explanation from information we receive from the world around us. When carrying out complex sensemaking tasks it can be important to maintain a record of the reasoning process. In contexts such as law and intelligence analysis it is imperative that a 'chain of custody' or 'paper trail' is maintained, keeping track of the control and analysis of data and information. This historical account of an analysis can help reduce uncertainty and increase trust in findings by allowing reasoning to be scrutinized, supports handover of analysis in collaborative settings, and can support the sensemaker's own understanding and confidence in their analysis. This historical account of reasoning in an analysis, known as its 'analytical provenance' [1], provides a description of the actions performed and techniques used at a given point in an analysis.

A. Wheat (✉) • S. Attfield • R. Fields
Interaction Design Centre, Middlesex University, London, UK
e-mail: a.wheat@mdx.ac.uk; s.attfield@mdx.ac.uk; b.fields@mdx.ac.uk

V.L. Lemieux (ed.), *Building Trust in Information*, Springer Proceedings
in Business and Economics, DOI 10.1007/978-3-319-40226-0_10

Although we have the ability to easily record actions and events in computer environments that can form part of an account of analytic provenance, this only paints part of the picture. An analysis does not take place in a vacuum, and sensemaking does not just take place in a person's head, but through elicitation and interaction with various artifacts and forms of representational media. Therefore an account of analytical provenance must include descriptions of the use and role of representational media in the sensemaking process leading to insights and findings.

In this paper we introduce the notion of *distributed sensemaking* and discuss how its concepts can help in creating a record of analytical provenance that includes an account of the role of representational media in sensemaking. Distributed sensemaking models the flow of information in and co-ordination of representational artifacts to form insights in sensemaking as 'inference trajectories', and provides a number of levels of description in characterizing representational artifacts.

The rest of this paper is structured as follows: in the next section we introduce distributed sensemaking, outlining its theoretical background before introducing the notion of inference trajectories and a number of levels of description. In Sect. 3 we discuss the concept of analytic provenance and the challenges in capturing a full account of analytic provenance including the use and role of representational media. We then go on to discuss how the concepts of distributed sensemaking can provide a foundation for research into the modeling and framing of analytic provenance in terms of the use of representational media and artifacts in sensemaking. Lastly we discuss challenges faced in capturing this type of provenance information across numerous types of media.

2 Distributed Sensemaking

2.1 Sensemaking

The term sensemaking literally refers to 'making sense'. For instance, sensemaking can take place when a holiday maker is comparing the best flight deals online, or when a detective is examining evidence in order to find the culprit responsible for committing a crime. When we engage in sensemaking, we embark on a process of comprehension [2] in which we seek out, re-structure and re-organize information in order to find meaning and construct a plausible understanding of some aspect of the world [3, 4]. Multiple theories of sensemaking have emerged—seemingly independently—in a number of research areas [5] including Organizational Studies [3], Information Science [6], Human-computer Interaction [7, 8] and Naturalistic Decision Making [2, 9].

Klein et al. [2, 9] offer a 'macrocognitive' theory of sensemaking involving the interaction of two types of entity: *data* and *frames*. Data are aspects of the world as experienced by the sensemaker through interaction with it. These might include things that a person might perceive in a given situation that may be important to them, such as a patient's symptoms in a medical setting or the co-ordinates and direction of aircraft in air traffic control.

A frame is a representation that accounts for the current understanding of something. For example, this could be the doctor's belief about the patient's medical condition, or it could be the air traffic controller's understanding of the flightpaths of aircraft in airspace he or she is responsible for. In this light, a frame serves as both an interpretation and explanation of the data available at a particular moment in time [10]. According to Klein et al., sensemaking is a continual process involving framing and re-framing when new data is available. As the sensemaker experiences a new situation, a frame acts as an interpretation of it. As more data becomes available, the current frame may be elaborated upon or challenged, causing the frame to evolve over time. As it does, it becomes a more plausible account of the situation as previous frames are rejected or modified in light of new data. Furthermore, as sensemaking is a bi-directional process, a frame may also call upon new data to be sought out, directing information seeking, and in so doing, revealing further data that changes the frame.

2.2 Distributed Cognition

Distributed cognition provides a perspective in which human cognition transcends the boundaries of the head of the individual, seeing intelligent processes as being distributed among people, the artifacts they use and the environment in which they are situated, and is affected by previous events and experiences [11].

In distributed cognition, cognitive activities are seen as computations that propagate representational state through a series of different media, which can occur both inside or outside of the head. For example this could be a person's memory, or external media such as charts or maps. The unit of analysis in distributed cognition, therefore, is a cognitive system which is made up of the internal processes of individuals interacting with a number of artifacts, each other and the environment in which they are situated. Studies of such cognitive systems include ship navigation [11], aircraft cockpits [12, 13], air traffic control [14] and emergency medical dispatch [15].

Hollan and colleagues [16] describe distributed cognition as three 'tenets': *socially distributed cognition*, which describes the distribution of cognitive tasks among individuals within a social group; *embodied cognition*, which describes the coordination between internal (the mind) and external (materials and environment) functions; and *culture and cognition*, describing how cognitive processes can be shaped by earlier experiences or social and cultural practices.

2.3 *Representational Media in Sensemaking*

Embodied within artifacts and representational media (e.g. maps, charts, lists) are a number of affordances which can furnish people with the ability to perform tasks that may otherwise be difficult to conduct solely in the head. These representations occurring 'in the world' are thought to change sensemaking in some way, but aren't addressed in much depth in existing sensemaking theory [10]. *Distributed sensemaking* addresses this by considering sensemaking through the lens of distributed cognition.

2.4 *Inference Trajectories*

In the distributed sensemaking paradigm, the flow of information throughout the sensemaking process across different representational media can be modeled as *inference trajectories*. An inference trajectory shows the relationship between information about some aspect of the world, extracted from representational artifacts, and its use in conjunction with information contained within other representational media (which could be internal or external). When used in conjunction with each other, these pieces of information (and the media they are contained within) lead to the generation of insights and a *situation picture*. A situation picture, similar to Klein's frame [9], is a sensemaker's current understanding of a given situation, representing a plausible picture of events taking place in the real world.

A situation picture can be represented either internally (in the head) or externally (in the world), embodied within some representational artifact. As a sensemaker gains more traction in their reasoning process, gaining more insight and understanding, the situation picture becomes clearer and more well defined.

Figure 1 illustrates an inference trajectory from the study of military signals intelligence analysis. The study was conducted on analysts within a military signals

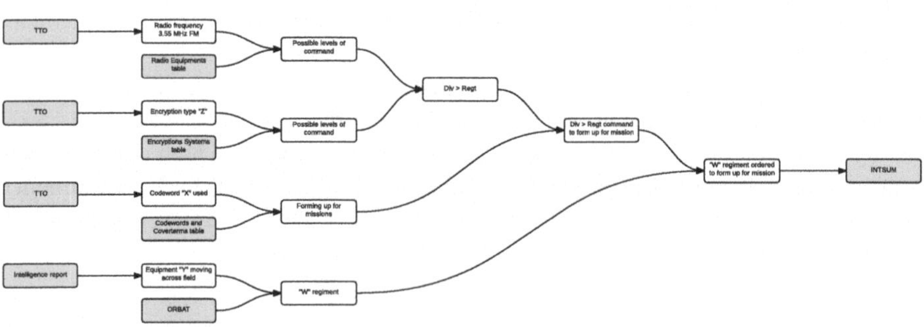

Fig. 1 An illustration of an inference trajectory showing the relationship and coordination of representational media in military signals intelligence analysis. Physical artifacts such as tables and charts are shown in *gray*. *White nodes* show information

intelligence cell, who's job it is to gain an understanding of the identity of enemy assets, their level in the command structure, their equipment and movements. In the study, analysts were fed extracts of intercepted radio communications in the form of 'tactical tip off' reports, or TTOs, using a number of 'working aids' to perform their analysis. These consisted of tables and charts containing known information about the opposing force, such as radio equipment information, known call signs, known use of code words, and intelligence about the command hierarchies and formations known as an Order of Battle (ORBAT). In the top-left (Fig. 1) the inference trajectory shows information extracted from a TTO—a radio frequency of 3.55 MHz FM—used in conjunction with a 'Radio Equipment table' leading to a number of possible levels of command. A similar operation is performed by the analyst revealing a further number of possible levels of command when the analyst uses information extracted from a TTO about an enemy asset using a radio encryption in conjunction with the 'Encryptions Systems table'. From this the analyst infers that the enemy level of command is 'Div → Regt'. This inference comes as the result of a boolean conjunction between the two lists of possible levels of command (for further explanation see [10]) and would be difficult to perform without the use of external representational media—in this case two tables of information.

2.5 Levels of Description in Distributed Sensemaking

Inference trajectories provide an abstract view of the information flow and co-ordination of representational media within the sensemaking process. However, the properties of artifacts leveraged by the sensemaker are also key to their use. For example, in the study described above, the analyst used a number of tables to carry out the sensemaking task. One such table was the 'Radio Equipments table', which contained known information about enemy radio equipment including frequency ranges, modes (FM,AM etc.) and levels of command within the military hierarchy which use certain frequencies. When working out possible levels of command of intercepted communications, the analyst would refer to the table and eliminate row by row—by striking through using a pen or pencil—those radio frequency ranges and types which the intercepted signal do not match. By doing this the analyst is deductively working out a list of possible levels of command. Moreover, as the analyst strikes out each row on the table, he reduces the number of possible levels of command for a signal, leading to an clearer situation picture.

We describe such properties at three distinct levels of description: *physical*, *semantic* and *pragmatic*.

Physical properties can be described in terms of an artifact's material and shape—how it is physically constituted. Moreover, when considering the physical makeup of an artifact, the affordances it offers in virtue of them are also considered. That is, the physical properties of an object which help, support, facilitate or enable physical action [17].

Semantic properties of artifacts are what they are taken to represent or stand for. That is, when used in sensemaking, artifacts are imbued with some representational meaning such that they represent some aspect of the world. For example, a database in a shop might represent a series of associations between products/stock levels and cost.

Pragmatic properties like the semantic properties of artifacts are concerned with their meaning and what they represent. However, where the semantic meaning of an artifact is constant, the pragmatic properties of artifacts are concerned with the role given to the artifact in current cognitive activity, which is subject to change. Namely, this is what an object is used for in virtue of its physical and semantic properties. For example, a shopping list might have items crossed or ticked off as they are put in the shopping trolley. To the shopper, this represents a list of items retrieved (crossed or ticked off) and items needed (not crossed or ticked off). Each time the shopper crosses off an item retrieved, the shopping list gains new meaning in terms of its cognitive role—it serves as an up-to-date record of items in the shopping trolley, and items not yet collected.

3 Analytic Provenance and Distributed Sensemaking

3.1 Analytic Provenance

An account of analytic provenance can be important in many situations, helping to reduce uncertainty and aid collaboration. Analytic provenance accounts for the actions and techniques used in an analysis at any point in time. In areas such as legal practice for example, it is important that a 'chain of custody' or 'paper trail' is preserved showing the control, transfer and analysis of evidence. By maintaining a record of analytic provenance, an account of the analytic process at any point is kept. This supports "reflection-on-action" [18] by allowing the interpretation and audit of claims and insights to be made, preserving a level of accountability and confidence in findings. Provenance information can also be important during an analysis itself by supporting "reflection-in-action", allowing people to interpret their own findings, identify areas in their analysis that might be weak and help them make sense of what they are trying to do [19]. Furthermore, in collaborative contexts—where analysts may be working as part of a large team or in non-colocated settings—an account of provenance can play a vital role in keeping track of individual actions which may not be clear from results alone [19]. This can be useful in assisting the coordination of labour, enabling best practice and supporting handover of tasks in an analysis.

3.2 *Provenance and Representational Media*

When sensemaking occurs, in many contexts reasoning takes place through the elicitation of a number of resources and representational media, both internally and externally. In Sect. 2 we discussed how we view this as distributed sensemaking. In light of that, it must be taken into consideration when recording analytic provenance, that there may be a variety of different sources of insight and knowledge. We previously introduced the study of military analysis which used a number of printed charts and tables known as 'working aids' alongside computer software and tools to generate insights and knowledge. The use of this type of representational media and external resources is commonplace, and as such, any account of analytic provenance may be seen as incomplete without a record of the flow of information and inference generation through the use of representational artifacts.

4 Challenges in Framing and Capturing Analytic Provenance

According to Xu et al. [19] there has been considerable progress in the capture and visualization of data provenance and analytic provenance, however, there is still some progress that can be made until it can be understood and used in terms of distributed sensemaking. Currently there is the ability within visual analytics systems to capture events such as mouse clicks, keystrokes and actions such as database queries and searches within computer environments [1, 19]. But this provides only part of the picture. As we have discussed, sensemaking occurs through the elicitation of different media, therefore to provide a full account of analytic provenance, we must find ways to capture it across the different representational media used within the sensemaking process. This presents a number of challenges. Firstly, the modeling and framing of this information requires further research to be carried out to have a more complete understanding of distributed sensemaking and to further develop a framework for its capture and analysis. Secondly, given the nature of different representations and artifacts—outside of the computer environment or inside the head—it is very difficult, if not impossible to automatically capture this information.

4.1 *Modelling and Framing Analytic Provenance in Distributed Sensemaking*

In Sect. 2 we introduced a model of distributed sensemaking including inference trajectories and a number of levels of description of representational artifacts. We believe this provides a foundation for research into the modeling and framing of analytic provenance in distributed sensemaking.

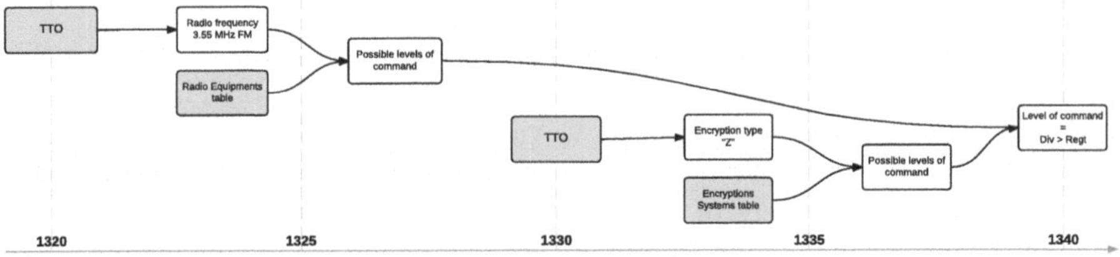

Fig. 2 A section of an inference trajectory showing the use of representational artifacts through time

4.1.1 Inference Trajectories

Inference trajectories show the relationship and coordination between information extracted from, and the use of different representational media in sensemaking scenarios. In Fig. 1 we have shown an illustration of an inference trajectory in military intelligence analysis. This provides a useful bird's eye view of an analyst's sensemaking (or analytic) process, showing how he or she has reached inferences and insights through the elicitation of different representational media. However, it does not provide a chronological account showing the development of the sensemaking process and flow of information over time—the analytic provenance. Figure 2 shows a section of the same inference trajectory which has been adapted to reveal the use of representational artifacts and generation of insights through time. By looking at a certain point in the chronology, a snapshot of the distributed sensemaking process—and the representational media involved—can be seen. Moreover, by viewing analytic provenance and the use of representation in this way, an account of events leading to insights and inference is visible in order of occurrence, allowing easy reflection on the analytic process and the status of information and knowledge at an point in time.

4.1.2 Levels of Description

Inference trajectories are a useful way of looking at the overall use of representational media and artifacts. However, this comes at a low level of resolution, and provides no detail about the make-up of artifacts or details of how they are used, which is important when reflecting on the use of representational media in an analysis or sensemaking task. In Sect. 2.5 we introduced a number of levels of description within the distributed sensemaking paradigm. These look at the *physical properties* of an artifact such as its material and shape and the physical affordances it offers; the *semantic properties* of an artifact, which look at the representational meaning given to an artifact; and the *pragmatic properties* of an artifact, which look at its role in current cognitive activity.

So, by describing the properties of representational artifacts at these levels throughout given points in the inference trajectory, we can show the use of representational artifacts at a given point in an analysis as well as how their use lead to insight and knowledge generation.

4.2 Capturing Analytic Provenance in Distributed Sensemaking

The capture of analytic provenance is a significant challenge. The capture of low level events and actions in digital environments is relatively easy [19]. However, this type of provenance information reveals only a limited picture of the sensemaking process, as much of this occurs outside of the computer environment across different physical media and inside the sensemaker's mind. The capture of analytic provenance therefore must occur, in part, manually. This however can be time consuming and labor intensive. Another issue is that of timeliness. It may be that the sensemaker may forget what they were doing at a given point, or what their thinking was when using a representational artifact, so without capturing provenance information within a limited timeframe, it could be lost or become less reliable.

There are contexts where analytic provenance—across different media—is already captured and forms an important part of maintaining reliability and trust in information. We previously mentioned this in the context of law, where a 'paper trail' of evidence must be preserved documenting the acquisition, control and analysis of evidence. In fields such as history, art and archival sciences a similar chronology of the status of artifacts must be maintained to determine authenticity. In these areas, the capture and recording of provenance information is already established, and may prove to be fruitful areas for research when facing the challenge of capturing and recording analytic provenance in distributed sensemaking scenarios. By conducting such research, we could learn efficient and well established methods of acquiring, documenting and preserving provenance information, which could be applied in the capture of analytic provenance.

5 Conclusion

Sensemaking does not occur only in the head of the individual, but through the elicitation of, and interaction with various forms of representational media. Therefore, a full account of analytic provenance in sensemaking scenarios must describe the use and role of the different representational media and representational artifacts in the process. This full account of the sensemaking process over time can be useful in a number of ways. It can support "reflection-on-action" by capturing points in the sensemaking process where resources are used together in order to reach insights, allowing the scrutiny and validation of findings, thus reducing uncertainty. Moreover, by reflecting on the use of representational media we can learn from the reasoning process, developing better materials and resources for sensemaking and analysis. It can also be a source of "reflection-in-action" allowing individuals to interpret their own findings and identify weaknesses in their analyses, as well as supporting collaboration by keeping track of individual actions.

In this paper we have shown how inference trajectories can keep track of the use of representational media over time, and in Fig. 2, we have illustrated this in a military intelligence analysis scenario. Also, by describing the physical, semantic and pragmatic properties of artifacts used at given points in sensemaking we can show how their use impacts on the sensemaking process.

There remains a number of challenges however. A key challenge is that of capturing analytic provenance in distributed sensemaking. Recording an account of analytic provenance which includes the use of representational artifacts must currently be done manually, which is time consuming and labor intensive. Also, there is a limited timeframe by which this information can be collected—a person may forget what they were doing or what they were thinking when using an artifact after a certain amount of time.

Looking ahead, we propose future research, including the study of distributed sensemaking in a number of scenarios, tracking reasoning and the use of representational media through the construction of inference trajectories. Here we can assess the utility of inference trajectories in these scenarios and further develop ways of modeling and framing analytic provenance in distributed sensemaking. We also propose research be carried out in finding more reliable and less costly methods for recording analytic provenance in distributed sensemaking contexts, for example by electronically tagging and tracking the use of artifacts in the environment, facilitating the automatic capture of their use through time.

References

1. Gotz, D., Zhou, M.X.: Characterizing users' visual analytic activity for insight provenance. Inf. Vis. **8**, 42–55 (2009)
2. Klein, G., Phillips, J.K., Rall, E.L., Peluso, D.A.: A data-frame theory of sensemaking. In: Expertise Out of Context. Proceedings of the Sixth International Conference on Naturalistic Decision Making, pp. 113–155. Psychology Press, Hove (2007)
3. Weick, K.E.: Sensemaking in Organizations, vol. 3. Sage, Beverly Hills, CA (1995)
4. Weick, K.E., Sutcliffe, K.M., Obstfeld, D.: Organizing and the process of sensemaking. Organ. Sci. **16**, 409–421 (2005)
5. Blandford, A., Attfield, S.: Interacting with information. Synth. Lect. Hum. Centered Inform. **3**, 1–99 (2010)
6. Dervin, B.: An overview of sense-making research: concepts, methods and results to date. In: International Communication Association Annual Meeting (1983)

7. Pirolli, P., Card, S.: Information foraging in information access environments. In: Proceedings of the SIGCHI Conference on Human Factors in Computing Systems. CHI '95, New York, pp. 51–58. ACM/Addison-Wesley, New York/Reading, MA (1995)
8. Russell, D.M., Stefik, M.J., Pirolli, P., Card, S.K.: The cost structure of sensemaking. Proceedings of the SIGCHI conference on Human factors in computing systems - CHI '93, pp. 269–276 (1993)
9. Klein, G., Moon, B., Hoffman, R.R.: Making sense of sensemaking 2: a macrocognitive model. IEEE Intell. Syst. **21**, 88–92 (2006)
10. Attfield, S., Fields, B., Wheat, A., Hutton, R.J.B., Nixon, J., Leggatt, A., Blackford, H.: Distributed sensemaking: a case study of military analysis. In: 12th International Conference on Naturalistic Decision Making (2015)
11. Hutchins, E.: Cognition in the Wild. MIT, Cambridge, MA (1995)
12. Hutchins, E.: How a cockpit remembers its speeds. Cogn. Sci. **19**, 265–288 (1995)
13. Hutchins, E., Klausen, T.: Distributed cognition in an airline cockpit. In: Cognition and Communication at Work, MIT, Cambridge, MA, pp. 15–34 (1996)
14. Halverson, C.A.: Inside the cognitive workplace: new technology and air traffic control. Ph.D. thesis, University of California, San Diego (1995)
15. Blandford, A., William Wong, B., Wong, B.L.W.: Situation awareness in emergency medical dispatch. Int. J. Hum. Comput. Stud. **61**, 421–452 (2004)
16. Hollan, J., Hutchins, E., Kirsh, D.: Distributed cognition: toward a new foundation for human-computer interaction research. ACM Trans. Comput. Hum. Interact. **7**, 174–196 (2000)
17. Hartson, R.: Cognitive, physical, sensory, and functional affordances in interaction design. Behav. Inform. Technol. **22**, 315–338 (2003)
18. Schon, D.A., DeSanctis, V.: The reflective practitioner: how professionals think in action. J. Contin. High. Educ. **34**, 29–30 (1986)
19. Xu, K., Attfield, S., Jankun-Kelly, T.J., Wheat, A., Nguyen, P.H., Selvaraj, N.: Analytic provenance for sensemaking: a research agenda. IEEE Comput. Graph. Appl. **35**, 56–64 (2015)

Appendix: List of Participants of the May 14–15, 2015 Workshop on Provenance: Past, Present and Future in Interdisciplinary Perspective, World Bank, Washington, DC

Lucie Burgess, Bodleian Library, Oxford, UK
Ken Cavelier, University of British Columbia, Vancouver, Canada
Adrian Cunningham, Queensland State Archives, Brisbane, Australia
David Dubin, University of Illinois-Urbana Champagne, Illinois, USA
Luciana Duranti, University of British Columbia, Vancouver, Canada
Larry Lannom, Corporation for National Research Initiatives (CNRI), Reston, VA, USA and Research Data Alliance
Victoria Lemieux, University of British Columbia, Vancouver, Canada
Bertram Ludäescher, University of Illinois-Urbana Champagne, Illinois, USA
Giovanni Michetti, Sapienza University of Rome, Rome, Italy
Paolo Missier, Newcastle University, Newcastle upon Tyne, UK
Corinne Rogers, University of British Columbia, Vancouver, Canada
Joe Tennis, University of Washington, Seattle Washington, USA
Ken Thibodeau, US National Records and Archives Administration (retired) and US National Institute of Standards and Technology, Washington, DC
Margaret Varga, Seetru Ltd., Bristol, UK and Oxford University, Oxford, UK
Ashley Wheat, Middlesex University, London, UK

V.L. Lemieux (ed.), *Building Trust in Information*, Springer Proceedings in Business and Economics, DOI 10.1007/978-3-319-40226-0

The manufacturer's authorised representative in the EU is Springer Nature Customer Service Centre GmbH, Europaplatz 3, 69115 Heidelberg, Germany. If you have any concerns regarding our products, please contact ProductSafety@springernature.com

Printed and bound by CPI Group (UK) Ltd, Croydon, CR0 4YY
25/09/2026
02222854-0020